BIG
GOOD

Philanthropy in the Age of
Big Data & Cognitive Computing

David M. Lawson

Edited by

Lori Hood Lawson

WORKINGPHILANTHROPY.COM, LLC
914 Railroad Avenue
Tallahassee, FL 32310

ISBN: 978-0-692-98369-0 (Paperback)

Library of Congress Control Number: 2017918120
WorkingPhilanthropy.com, LLC, Tallahassee, FL

First Printing: December 2017

Cover Design by Galen Lawson
Illustrations by Max Lawson

While the debate rages over whether Big Data and Cognitive Computing are going to save or destroy our way of life, or even perhaps life itself, most non-governmental organizations are on the sidelines waiting to see who wins. In the meantime, we have never been faced with more urgent, and complex, problems needing solutions now. From climate change to homelessness to helping students turn into successful alums, the answers are not going to be found on traditional business intelligence dashboards or by using old-school analytics based on incomplete data. This book is for the people who dedicate their time, talent, and treasure to help others, and who need actionable insights to further their organization's mission.

It is dedicated to my mom, who never stopped believing in me - I miss you every day; to my dad, who has used his talents for good; and to Lori, who inspires me to be a better person (and writer). This book would not have been written without you.

Table of Contents

Preface

This book is not aimed at my fellow lovers of data and technology (though I do believe they will benefit from it). It is for the people who will put the actionable insights the data and technology yields to practical use. For that reason, the book is meant to provide a working understanding of the concepts, not a deep dive (for that, look at the "So You Want to Know More" section at the end). Most importantly, it is intended to provide you with real-world use cases, which I hope will spark your imagination.

I have three professional passions: philanthropy; data; and technology. These have guided my career, from the first time I conducted research on private foundations as a teenager in the 70s to my latest venture, NewSci, bringing Cognitive Computing to our field. In between, I have been privileged to work with literally thousands of organizations, from the ones you all know to small local charities known only to their local community.

I did not fall far from the family tree: my grandmother was a key punch operator for the railroad, as well as lay leader in her church; my father is one of the pioneers in fundraising who also owned one of the early fax machines (it actually burned the words into the paper); my mother was a fundraiser, and she helped to introduce Lanier® word processing machines (and later IBM® PCs) to law offices.

My fascination with technology began with a simple game of Pong at the Nashville airport in 1974. My mom and I played for an hour while waiting for our flight. When she started installing Lanier word processors at a DC law firm a few years later, I went with her and realized very quickly this was going to replace the IBM Selectric® typewriter I had been using.

Being able to not only write research profiles, but also store them in something other than a file cabinet, was transformational. Then, on-line look-up services came out, and I was able to integrate them with my library research. Seemingly overnight, I was able to scale a research business.

During this time I also encountered a challenge, which I would face until Big Data and Cognitive Computing came along — how to search and analyze the documents I was creating. Back then I used a program called ZyIndex®. Years later I found out it was also being used by the FBI to search documents. While it was great at retrieving data based on Boolean searches (AND, OR, etc.), there was no way to analyze it.

My experience is why when Big Data came along I understood it right away. The ability to store both structured (fielded) and unstructured (un-fielded) data in the same place was a dream come true. My goal with this book is to get you and your team as excited about this new technology as I am.

I have been fortunate (and at times vexed) to be involved in some of the transformational moments in our industry. In the late 80s, I brought asset screening based on stock holdings of corporate insiders to fundraisers. I learned a valuable lesson: just because something is really valuable to you doesn't mean others will see the value, at least not right away.

The first response was, "I don't have those people in my database." This was followed by, "Why would I want to know about their stock holdings." I was able to answer the first question by running tests. The second was more involved because I had to start Wall Street for Fundraisers seminars to explain the complexities of stock ownership, and how stock could be donated to charities. Ultimately, it was not until a client raised a million dollars from a prospect we had found for their organization that wealth screening really took off.

In my next company, Prospect Information Network (P!N), I brought to our sector a networked database for the management of wealth screening results – ProfileBuilder™. This was met with, "We can't have a shadow database." I had to point out their statement assumed there was a place to store and manage screening data in their current database. At that time, there was not.

ProfileBuilder also integrated access to the world-wide-web, which was quickly filling up with information sources useful for prospect research, and much of it was free. This was met with, "We can't trust data from the web!"

Over the years I have done a lot of presentations, and with the Council for Advancement and Support of Education (CASE) I worked on some of the first on-line fundraising conferences. This was met with, "People won't give on-line," and "Fundraising is a personal business."

Even email was fought by some senior leaders in the field until early in the 2000s (and some still fight it to this day). They just could not get their heads around the idea that people were starting to use email instead of letters. The consensus among these leaders was an email is too impersonal to be treated the same as a letter.

Then I sold P!N to Kintera, and entered the world of crowdfunding and Software-as-a-Service. Here, people would say, "No one is going to put their data outside of their organization," and, "Online giving is never going to be a major part of fundraising."

You can see a trend here. When faced with something new, the first reaction is to try to stop it in its tracks. Remember when people said social networks were a fad?

I have spent my career going around the corner to see what is up ahead and bringing it back to the philanthropic community. I feel strongly we deserve the very best tools for this simple reason - we do such important work. I invite you to come with me as we look at what is shaping up to be the biggest technological transformation of our time. There are plenty of reasons not to embrace Cognitive Computing and Big Data. A few years from now, all of these objections will be as unfounded as "People don't give online" is today.

01

The Good Race

"Life is neither good or evil, but only a place for good and evil."

—Marcus Aurelius

©2017 Max Lawson

Until Cognitive Computing, technology has been a *tool* for humans to do good or evil. Now, it can *do* good or evil on its own. This is what truly scares us as we contemplate artificial intelligence becoming real intelligence. It is not that it will be more than us. It is that it will be just like us.

The battle between good and evil is as old as our human race is. It is the eternal struggle to tamp down the worst of our instincts while at the same time heightening our ability to make the most of our best selves.

Technology has mirrored this same battle. The computer enabled us to land on the moon, yet it also enabled us to create the atom bomb. The Internet has brought the world together, and also helped people terrorize us.

So what are we to do? Stop developing the technology? Create cognitive disarmament treaties? First, it is not practical to stop. Second, it would never stop anyway. The technology is too pervasive, and I would argue it is too important to our future.

Rather than a battle, this will be a race between the good and evil *uses* of the technology. A major goal of writing this book is to help encourage the good guys to pick up the pace. Let's not turn this into the same debates we had over the value of putting data into databases in the 80s; whether email was *personal* enough in the 90s; whether the world wide web was more than a place to track your stocks or visit XXX.com in the 90s; whether people would give on-line in the 2000s; and whether social was a fad focused on the young – a debate I still hear to this day.

Too much time was lost in our sector as these debates raged. The instinct to wait until new ideas have become old has cost not only money, but also actual lives.

- How much efficiency was lost as organizations waited to adopt databases?
- How many donors were not engaged because they were not contacted via their preferred channel of email?
- How many potential members and donors didn't see your organization on the web while shopping and researching for-profit companies because your website was just a brochure?
- How many people didn't you reach on social media because you assumed your donors were not there?

Five years from now will you look back and wonder how many insights you missed because you waited for Cognitive Computing to be what everyone was doing?

While Cognitive Computing will yield enormous gains for your fundraising, what should really get you moving is what it can do for your mission. Cognitive Computing will ultimately be measured by how it enabled your mission to deliver greater impact, not by how much more money it helped you raise (although it will do that too).

Fight your natural instincts to make do with what you have. This is noble and is embodied by the millions of volunteers, and the passionate low-paid staff. It is almost with pride people in our sector speak of having to do more with less.

I call foul on this argument. It is time to do more with more.

Our work is far too important to be treated like a second-hand-sector. Education, healthcare, faith, environment, quality of life, and the future of society don't sound like second-tier concepts.

The cognitive age will be a time of unimaginable discovery. Medical breakthroughs will become commonplace as will new ways of learning. Seemingly insurmountable environmental problems will be solved in ways we cannot even imagine today. Fresh vegetables will be grown down the street, not across the ocean, in effect reducing transportation costs to almost zero while also delivering produce with real taste.

These are not utopian dreams. All that has ever stood between us and progress was imagination, and the ability of a few humans to transcend the status quo. In the cognitive world, our imaginations can be turned into reality in days not years. The whole idea of the status quo becomes obsolete as everything is always challenged.

We also don't have to wait for an Einstein or Jobs to show us the way. We will have 10 billion people on this planet in the not too distant future. We can't take the chance Einstein decides to just make gadgets instead of focusing on theory or that Jobs decides to live in a commune instead of bringing the first consumer-centered technology to the world.

In this race, companies seeking to make a profit are not necessarily evil. In fact, most companies are doing plenty of good by making things we need and employing people. The evil in the cognitive era will be those individuals and companies choosing to harness its capabilities to harm people and the planet.

This is no different than a company using the phone for a scam or using a computer to hack into another company's data to steal it. We sadly know all too well planes and cars can be used as weapons.

A sure way to stop all of it would be to take away all the computers, phones, planes, and cars. But then you would also take away an organization's ability to reach out to mission beneficiaries or for passionate supporters to reach out to the organization. How would disaster relief reach far off places? Will everyone walk or bike to work? (That one I'm actually okay with.)

Rather than round up all the Cognitive Computing, let's put it to the best possible uses. At the same time let's establish cognitive regulations. People using a phone for a scam are doing something illegal, and, in the United States, we have the Federal Communication Commission to monitor such. You can't just buy a plane (or even a drone) and fly it around. The Federal Aviation Administration issues certificates and has maintenance requirements. Along with your license from the DMV, your car has to be inspected, in some states, and, at some point, an officer is going to ticket you for not fixing your brake light or for running a red light.

In the chapter on Data Governance, I talk about the need for a well thought-out plan for not just gathering but also governing data and insights. Organizations need to establish clear policies regarding their use of Cognitive Computing. We are likely to see cognitive regulations at the local, State, Federal, and global levels. The General Data Protection Regulation (GDPR) has already woken up organizations in the European Union to the need for strong Data Governance.

Just like any sanctioned race, there are going to be rules. You won't be able to go anywhere the technology takes you. Cheating will not be tolerated. The rules will not always seem fair, and will more than likely favor some over others. This is just the way the world is.

My strong suggestion to the associations in our sector is to get out front. Don't sit back and wait for policies and regulations to be forced on you. Help government shape the path to the future, and help your members navigate it.

When HIPAA (Health Insurance Portability and Accountability Act) came along, it was frustrating how many nonprofit hospitals were either unprepared or just felt victimized by the Act. Only recently have the rules reflected the reality of these hospitals needing patients to give in order to provide them the level of service they expect. We cannot afford to fix harmful cognitive regulations after they are adopted. The time to act is during their formation. If you are not at the table, then you risk being a victim.

The companies who stand to benefit from the adoption of Cognitive Computing understand if they don't address the concerns of society, they may be stopped (or at least slowed) from developing it. To that end they made the following announcement:

> September 28, 2016

> NEW YORK — Amazon, DeepMind/Google, Facebook, IBM, and Microsoft today announced that they will create a non-profit organization that will work to advance public understanding of artificial intelligence technologies (AI) and formulate best practices on the challenges and opportunities within the field. Academics, non-profits, and specialists in policy and ethics will be invited to join the Board of the organization, named the Partnership on Artificial Intelligence to Benefit People and Society (Partnership on AI).

> There are seven initial pillars of the organization:

>> **Safety – Critical AI** - Advances in AI have the potential to improve outcomes, enhance quality, and reduce costs in such safety-critical areas as healthcare and transportation. Effective and

careful applications of pattern recognition, automated decision making, and robotic systems show promise for enhancing the quality of life and preventing thousands of needless deaths.

However, where AI tools are used to supplement or replace human decision-making, we must be sure that they are safe, trustworthy, and aligned with the ethics and preferences of people who are influenced by their actions.

We will pursue studies and best practices around the fielding of AI in safety-critical application areas.

Fair, Transparent, and Accountable AI - AI has the potential to provide societal value by recognizing patterns and drawing inferences from large amounts of data. Data can be harnessed to develop useful diagnostic systems and recommendation engines and to support people in making breakthroughs in such areas as biomedicine, public health, safety, criminal justice, education, and sustainability.

While such results promise to provide great value, we need to be sensitive to the possibility that there are hidden assumptions and biases in data, and therefore in the systems built from that data. This can lead to actions and recommendations that replicate those biases and suffer from serious blind spots.

Researchers, officials, and the public should be sensitive to these possibilities and we should seek to develop methods that detect and correct those errors and biases, not replicate them. We also need to work to develop systems that can explain the rationale for inferences.

We will pursue opportunities to develop best practices around the development and fielding of fair, explainable, and accountable AI systems.

Collaborations Between People and AI Systems - A promising area of AI is the design of systems that augment the perception, cognition, and problem-solving abilities of people. Examples include the use of AI technologies to help physicians make more timely and accurate diagnoses and assistance provided to drivers of cars to help them to avoid dangerous situations and crashes.

Opportunities for R&D and for the development of best practices on AI-human collaboration include methods that provide people with clarity about the understandings and confidence that AI systems have about situations, means for coordinating human and AI contributions to problem-solving, and enabling AI systems to work with people to resolve uncertainties about human goals.

AI, Labor, and the Economy - AI advances will undoubtedly have multiple influences on the distribution of jobs and nature of work. While advances promise to inject great value into the economy, they can also be the source of disruptions as new kinds of work are created and other types of work become less needed due to automation.

12

Discussions are rising on the best approaches to minimizing potential disruptions, making sure that the fruits of AI advances are widely shared and competition and innovation are encouraged and not stifled. We seek to study and understand best paths forward and play a role in this discussion.

Social and Societal Influences of AI - AI advances will touch people and society in numerous ways, including potential influences on privacy, democracy, criminal justice, and human rights. For example, while technologies that personalize information and that assist people with recommendations can provide people with valuable assistance, they could also inadvertently or deliberately manipulate people and influence opinions.

We seek to promote thoughtful collaboration and open dialogue about the potential subtle and salient influences of AI on people and society.

AI and Social Good - AI offers great potential for promoting the public good, for example in the realms of education, housing, public health, and sustainability. We see great value in collabo-rating with public and private organizations, including academia, scientific societies, NGOs, social entrepreneurs, and interested private citizens to promote discussions and catalyze efforts to address society's most pressing challenges.

Some of these projects may address deep societal challenges and will be moonshots –

ambitious big bets that could have far-reaching impacts. Others may be creative ideas that could quickly produce positive results by harnessing AI advances.

Special Initiatives - Beyond the specified thematic pillars, we also seek to convene and support projects that resonate with the tenets of our organization. We are particularly interested in supporting people and organizations that can benefit from the Partnership's diverse range of stakeholders.

We are open-minded about the forms that these efforts will take.

These pillars make it clear how intertwined with society AI will become. This intersection is where organizations have the opportunity to re-imagine themselves. The Partnership for AI sees this, and has actively recruited non-profit organizations to be part of its leadership. There also will be no shortage of opportunities for organizations to enlist the help of companies, and experts, to work on the challenges their missions are addressing.

It would be easy to rely only on companies and organizations serving technical and data professionals to help us manage the cognitive era, but it would be a potentially crippling mistake. Without the doctors, nurses, lawyers, business executives, entrepreneurs, farmers, fundraisers, marketers, and all the others who can act on the insights there will be no Return on Investment (ROI).

Yes, there are many ways to use insights derived from Cognitive Computing without a human involved, but those alone will not be able to realize the full potential of this technology. Ideally, professional associations of all types will work together to bring forward comprehensive plans to fully utilize Cognitive Computing, while also ensuring any negative impacts are mitigated.

So gather up all your good ideas, suppress your urge to fear the future, and enter the race to find the good places for us to go.

02

Meeting Watson

"Once an organization loses its spirit of pioneering and rests on its early work, its progress stops."

—Thomas J. Watson

The first technology, other than a game, that really captured my imagination was the Apple® Macintosh. In 1984, I went to a hotel in Washington, D.C. to behold the miracle of a computer with a graphically rich screen, and a device for clicking on what you wanted to do. The second time this happened was in the early 90s at the development office of a university in New Orleans where I saw words and graphics slowly (very slowly) appear on a screen thanks to the world-wide-web.

In both cases, I was inspired to rethink how I would use technology going forward. In neither case did I think of the technology as anything more than a tool. Then came IBM Watson.

Like most everyone else, my first introduction to Watson was hearing of him winning *Jeopardy*! A few years later I co-founded a company, NewSci, to bring the underlying technology of Big Data and Cognitive Computing to the philanthropic community.

Then in 2014, I had the opportunity to submit a design for an application built on the Watson platform itself. Only a handful of companies outside of IBM had been allowed to use the platform, so it was an incredible honor when the submission was accepted. We had only a few months to build our prototype as the new Watson HQ in Manhattan was to open in October, and we had an opportunity to be there if we were ready.

Documentation was the first clue of the magnitude of the platform. Watson could ingest any amount of data (200 million documents for *Jeopardy*!), but it had to be formatted a certain way, and while it was skilled at unstructured data it couldn't handle traditional structured fields.

It became clear we would need to form a teacher-student relationship, and then a supervisor-employee one, and at some point, I would need to trust it enough to take direction from it. The student becomes the teacher. I am hearing "David, I am your teacher" as I write this.

This idea of forming a relationship with technology rather than seeing it as a servant to do our bidding is exciting and unnerving. We have seen this in movies for decades but thought it was always going to be in the future. We have caught up with that future and now live in it.

IBM understood this and gave it the name of a person, not some techie name like Deep Blue, which became a chess champion. Many people initially believed it was named for Dr. Watson of Sherlock Holmes fame. It was in fact named after Thomas Watson, the first CEO of IBM, who I had referenced many times over the years because of his belief 100 was the perfect number of relationship-sales prospects in a salesperson's portfolio. While our field often has hundreds of prospects in a major gift officer's portfolio, I believe only 100 of them are being managed effectively

During a talk where I referenced Thomas Watson's approach to relationship sales, one of his granddaughters happened to be in the room. She came up to me after the presentation and said he would have liked what I said. I was grateful and relieved. For philanthropy, it is a good sign Watson was named for a person who understood the value of relationships.

NewSci's Watson app, Impact Measurement and Analysis (IMA), was approved and we were invited to exhibit at the opening of the Watson HQ. Arriving at the building I was struck by how much it felt (and even looked) like the business incubator I co-founded in Tallahassee, Domi Station. Despite IBM being the oldest tech company on the planet, that day it felt like a start-up (or at least a start-up with a billion dollar initial investment).

Watson was introduced to the group in a surround theater where we watched as he diagnosed a sick child exponentially faster than human doctors had in the real world. There were many factors, but ultimately the key was Watson taking into account the child's red eyes, which had been dismissed by doctors who thought it was caused by his crying. In fact, red eyes in conjunction with his other symptoms meant he had Kawasaki Disease.

Even very smart doctors have their confirmation biases, and sometimes those can literally cost lives. It was a reminder of AI's power to see the world as it is rather than as we as humans perceive it to be.

The next way we got to know Watson was over a meal. Watson had created recipes for the treats we enjoyed. They were very tasty, but as you looked at the ingredients you realized they were also very expensive. It was clear Watson's job was to create imaginative dishes without regard to cost.

As I enjoyed the food I was realizing the insights Watson shared needed to be combined with the realities of the organizations using them. In the case of recipes, Watson was ready to serve high-end restaurants, but would need new training if it was to deliver low-cost food for the hungry.

The main event was on an unfinished floor of a 12-story building in the heart of New York City's Silicon Alley. Don't think Fifth Avenue or Wall Street. It is less than a 15-minute walk from The Bowery Mission. IBM was making a statement with the location. It was a new beginning for them as much as it was the dawn of the Cognitive Computing age.

Talking with the other app developers, the press, and even IBM executives, it was apparent no one was completely sure what all of this would mean. IBM may have invented the underlying technology, as they had done with relational databases decades prior, yet they were still wondering what it would all become.

I wasn't worried because this is how all truly new inventions begin. No one, including the inventor, fully knows how it will be applied to current problems much less future ones. It is for this very reason IBM had invited entrepreneurs like me to be part of the launch. They need all of our imaginations working on the possibilities.

As I went from booth to booth to see what my fellow dreamers had come up with, I was struck by how many aspects of life were already being re-imagined. The founder of Travelocity® had created WayBlazer® to provide a cognitive travel service capable of delivering not just lists of hotels and attractions, but a complete itinerary for your trip based on your unique interests and desires. Cyber security and pet health were also represented.

Watson had begun its move beyond games and healthcare into the mainstream of society. Cognitive Computing could now be part of everyday life. We just had to imagine it.

The more I spoke with people the more excited I became, yet I also had a growing sense of the challenges ahead. Watson was unlike any technology we have ever encountered. We didn't just have to learn how to use it, we needed to get to know it.

To harness its power we would need to be teachers, not programmers. Subject matter expertise would be critical to successful implementations. Technical talent alone could not deliver a cognitive application.

Another revelation was Watson of *Jeopardy!* fame is really Watson One. None of us were using Watson One - we were using offspring with plenty of his digital DNA, but it was not the same Watson who had bested Ken Jennings. Each of the offspring had its own personality – ours was focused on mission impact while WayBlazer's was all about the travel experience.

You might think this is just like the different databases created from Oracle or SQL for various industries, but it's more than configuring the technology to meet industry specifications. It is about developing the capacity to understand a problem complete with the knowledge and perspective necessary to solve it.

As I write this, stories are appearing about the failure of Watson. The press, feeling they had been beguiled by all the hype, is striking back with tales of how the technology has not been able to cure cancer, and companies have struggled to fully realize its promise in other sectors.

The theme of these stories is, "If Watson is so smart, why can't it......?"

I am reminded of my favorite commencement speech, given by Conan O'Brien of late-night fame. At Harvard, he discusses a challenge of being successful:

> "Success is a lot like a bright white tuxedo. You feel terrific when you get it, but then you're desperately afraid of getting it dirty, of spoiling it."

I envision Watson wearing his white tuxedo after the *Jeopardy!* win, looking at everyone expecting him to now be able to do anything. Does he risk spoiling the impression he knows everything by attempting something new?

Fortunately for us he did, and he has gotten his tux dirty. He, and his offspring both inside and outside of IBM, have also begun to revolutionize the analysis of mammograms; deliver 24/7/365 customer support; improve crop yields by detecting pest infestations from photos; and enabled my company's work to improve 911 services among thousands of applications across the globe.

We, and the press, will have to adapt to having an iterative technology, one needing to fail before it succeeds. Imagine if someone had followed Einstein's every step in his journey to discovery. At every point, except the last, he would have been deemed a failure.

Harnessing this power will take imagination and patience. It will surely fail us at times, and may even imperil us. If we are honest, we accept failures every day, and we face ever-growing perils, so we are risking little to welcome this new technology into our lives and help it become a valued partner in our journey.

And don't forget the first Macintosh bombed. It was beautifully designed, but underpowered. I would not be writing this book on a MacBook Pro® if Steve Jobs had not dared to get his white tux dirty (maybe that's why he wore a black t-shirt), and more importantly didn't listen to those who said because he had failed, he could never succeed.

03

Big Data – Your Insight Reservoir

"It is by a thorough knowledge of the whole subject that our fellow-citizens are enabled to judge correctly of the past and to give a proper direction to the future."

—James Monroe

I wish we could retire the word data. I have spent my career seeking it, storing it, managing it, and trying to create value from it. What I have come to realize is most people care about data in much the same way they care about gold. People appreciate it has value, but don't want to know how it was extracted and made useable. Not a moment is spent pondering how gold is protected. What people care about is the ring on their finger (or that of their loved one).

Data's equivalent of jewelry is insight. Those are the jewels of data we are seeking.

We are not going to cover the many reasons you should care about how the gold in your jewelry was sourced, such as the working conditions in the mines, and how (or if) people are being paid. I suggest you visit Brilliant Earth to learn more - https://www.brilliantearth.com/gold-mining-labor-concerns/.

We are going to cover how to ethically source your data in the chapter on Data Governance. Ethics not only matter in the cognitive age, but matter more than ever.

Let's get back to retiring the word data.

One of the reasons for the slow pace of Big Data adoption is the impression it will just be a bigger warehouse to store data. Lots of talk about security, but that is as interesting to leadership as Fort Knox is to a jewelry shopper.

You will see the word insight a lot in this book, and that is purposeful. Data Mining becomes *Insight Extraction*. Data Analytics becomes *Insight Learning*. Data Visualization becomes *Insight Envisioning*. Data Warehouse becomes *Insight Reservoir*.

As much as I want to get onto insights, we must pay our respects to data. While you may not want to know how the insights are made, without a working knowledge you run the risk of not being able to act on the insights.

Before you skip ahead, Big Data, unlike the little data we have been using, has a depth and richness able to capture not just your attention, but also your imagination. We will look-back at the data we have relied on to make decisions as we now think of a pay-phone while using our smart-phone.

Big Data — What is it? Why should I care? How the heck am I going to use it?

Those three questions will guide our journey together. If my time bringing new technology and data to the philanthropic community has taught me anything, it has taught (no drilled into) me this: if people don't understand what you are talking about they may smile, but they will not follow you.

Once you understand what the new thing is, the real work begins, explaining why you should care when there is no shortage of things to do with the technology and data you already have. This is when classic objections arise, like, "We already do that," or "We can't do something like that here."

Finally, having answered the first two questions you are faced with the most difficult. What are you going to do with it? And perhaps more importantly, how is it going to change your organization for the better?

Let's start with Big Data.

Storage of data has been a problem from the beginning of the computer age. My first computer, an IBM PC, and its two floppy disks, had 160K of storage. That's not a typo – 160K, not 160MB or GB. 160,000 precious bytes of data. You can imagine how happy I was when the 3.5-inch disk was introduced with a whopping 1.44MB.

What this meant was databases were designed to efficiently store data, which translated into a lot of data being abbreviated (two letters for states; and two numbers for the year). Leaving out the 19 in the year led to the millennium bug where computers literally couldn't handle the new century. One of the fixes was to just change all birthdates before a certain year to start with 20 to buy time as we entered the 21st century.

When the world-wide-web came along, the need for storage grew exponentially as the amount of data exploded, and the types of data expanded to include images, video, and audio. This all came to a head as companies like Yahoo!® and Google® wanted to store and search on all the data on the planet. They literally brought the old relational databases to their knees.

The pressure of all of us wanting answers (and fast), and companies wanting to provide them, created a whole new way to store data – Hadoop®. Where did this name come from? It was the name of the stuffed elephant belonging to one of the Hadoop creator's son.

What Hadoop did was create a way to store any kind of data, in large amounts, on a distributed set of computers without fields. This was combined with MapReduce where each piece of data was reduced to just one element, connected to all other places where that element exists. For example, my name is very common, David, yet within this environment when I search for David it only looks in one place and then from there points to the millions of other places I and my fellow Davids exist.

This also meant when you bring data into Hadoop, unlike with a data warehouse, you don't map the data - you index it. This is much faster and doesn't rely on organizations having to go through the painful task of figuring out where all their data should go on the new platform. It is not hyperbole to say you can be up and running within a few months. Contrast this with the often year to two-year traditional data warehouse project, and you can see why Hadoop adoption is on a hockey-stick trajectory.

When a concept like Big Data comes along, one of the challenges is explaining it in a way which adds clarity rather than confusion. Since it is obviously big, experts decided to break it into its parts.

V was the letter chosen by experts to describe the components of Big Data. What got this started was the first V, which is the most obvious – *Volume*. This is simply the amount of data you are storing. Unfortunately, this is often seen as the determining factor in deciding whether you should embrace Big Data or not. It is easy to say, "We are not Google or Walmart®," and then think Big Data is not going to help your organization.

It is the second V that you should focus on – *Variety*. Is your data housed in a variety of locations, in a variety of formats? If the answer is yes, then Big Data is for you. There are three types of data:

> **Structured**: This is the data in traditional databases with fields (first name, middle name, last name, etc.)

> **Semi-Structured**: This data is not in fields, but has tags to describe it. XML and JSON are examples.

> **Unstructured**: This is 80-90% of data found in PDF; MS-Word; video; audio as well as social media posts and emails.

The third V is about how fast your data changes – *Velocity*. How often does your data change? Daily? Hourly? Minute-by-minute? Second-by-second?

Think about this: some experts have said driverless cars will produce 1GB of data a minute. You most likely are not in the driverless car business, but you may well have data that is changing every day, which is critical to understanding where your organization is heading.

The fourth V focuses on how reliable the data is – *Veracity*. What is the quality of your data? How messy is it? This V comes down to how much you can trust the data.

With Big Data you have to balance Veracity with the final V (at least for now) – *Value*. This is the value of the insights you are gleaning from the data. Just as it was with old-school databases, it was not until the value of having your data stored, and managed, exceeded the cost that it was fully embraced.

Deciding whether you should embrace Big Data or not begins by taking an inventory of your internal and external data repositories. Think cross-organizational, not just your own department. If you are a school, it means from prospective student marketing to admissions to students to alumni to faculty to volunteers to boosters. For healthcare, it is the community, patients, volunteers, physicians, and donors. For non-profits, it is the community, partners, members, volunteers, and donors.

As you explore your data landscape don't miss small niche databases and even spreadsheets being used to manage aspects of your organization. Also look at any external data such as wealth screening. Don't forget your virtual world data, including websites and social networks.

When you are done, you are likely going to be surprised at just how much data (Volume) you have, and how many formats (Variety) it is in.

Armed with the inventory, you can look at how often the data changes (Velocity). Here again, you may be surprised at how much data is flowing into your organization on a daily basis.

As you build your inventory, look at the quality of the data (Veracity). This will be subjective at this point (we will explore

Data Governance in a later chapter), but a simple criterion of good, okay, and poor is fine. Your goal at this stage is to gain an overall picture of your data.

The final step is the most important – identifying the actionable insights (Value) you could gain if you implement Big Data. This is not about knowing all the insights you will gain. In fact, as we will discuss later, the key to success is not thinking you know all the answers. What you want is to look at critical organizational questions and goals which deeper insight could help.

Why are some beneficiaries of our mission more successful than others? What causes a donor to stop giving? What drives engagement? What drives disengagement? What aspects of our mission are working the best? What aspects of our mission are not working? How can we raise more money?

You may have heard of the term *Data Lake*. This is a description of the platform where all the data is housed. While I understand why a lake was chosen as the metaphor, given many streams of data flow into it, I prefer *Insight Reservoir*.

An Insight Reservoir captures two critical components of a successful Big Data implementation: first, it puts the value on the insight, not the data; and second, a reservoir is owned and managed by your organization while a lake implies an openness which should not exist.

Your Insight Reservoir will collect, store, manage, transform, and distribute data as actionable insights to end-users. If you are more comfortable with Data Reservoir or even Data Lake, that's fine. I ask two things of you: first, make it Governed Data Reservoir or Governed Data Lake; and second, don't forget it will be actionable insights you are judged on, and, unlike a lake, it is not open to the public.

Before we move onto Cognitive Computing, I will share an example of how Big Data can avert disaster. In this case, perhaps the most famous disaster of them all – the sinking of the Titanic.

A new theory emerged a few years ago about why the Titanic captain steamed at full speed into an iceberg even though he had been warned the ship was heading into treacherous waters. The captain had traversed the Atlantic for decades, and in fact, this was scheduled to be his final voyage. He knew the waters as well as anyone.

His experience may have proved his undoing. When he received information from the telegraph office that other ships were reporting large iceberg fields, he may have dismissed them as the ranting's of less experienced captains. In his experience at this time of year, there were not icebergs, at least not in large numbers.

What he didn't know was there had been a celestial event in early January of that year which created a super moon, which had not happened for 1,400 years. That tide literally lifted icebergs over the Greenland fjords. The captain would have been right except for this year. The next time a captain may be wrong is the year 2257.

The Big Data lesson is if weather data had been analyzed as it is today, the captain would have known icebergs were going to be more prevalent than usual, so the messages may not have gone unheeded. He would have slowed down, and we would have been spared all of the movies, and, with all due respect to Celine Dion, that song.

Are there aspects of your organization you are 100% sure of because of how things have always been? Next time you say that, remember the Titanic, and know the past doesn't predict the future if an important variable changes.

Dark Data – What You Can't See Is Costing You

We don't need another scary term, but Dark Data is something you need to know about. From notes in your CRM to comments on your Facebook page to what is being said in audio recordings of your call center, this is all of the digital data you are not analyzing.

IBM has estimated 80% of data is dark, and this will rise to 93% by 2020. Imagine only analyzing 7% of your data. You are considered legally blind when your field of vision is 20% or less, so it should be no wonder organizations are struggling to find their way forward.

The first step to bringing your Dark Data into the light is to go on a data hunt. Keep in mind, while a lot of Dark Data is unstructured, not all of it is. You are safe to assume you have plenty of unexamined data in the rows and columns of your databases.

DATA TYPE

SOURCES	Structured	Unstructured
External	• Wealth ratings • Constituent scores • Asset details • Social media metrics • Social location/time • Charity ratings • Patient ratings • Review ratings	• Social media posts/comments • Sector white Papers/publications • Organization reviews • Employee reviews • Constituent comments • Constituent videos • Constituent images • Peer organization websites
Internal	• Internal ratings • Constituent data • Giving history • Survey ratings • Organization metrics • Prospect management data • Web logs • Social media metrics • Communication metrics • KPI metrics	• Verbatim Survey Responses • Organization videos • Class notes • Call center audio • Prospect call reports • Note fields in databases • Research profiles • Field reports • Organization Websites • Mission impact reports • Emails, letters & texts • Direct marketing copy

Once you know where the Dark Data is located, begin to connect it to your organization's operations and goals. What you will find is it often provides the *whys* to go with your business intelligence *whats*. Perhaps your key performance indicator related to donor retention has been falling. Dark Data may provide insights into why donors are turning away from your mission through their verbatim answers to survey questions; call reports from gift officers; and/or tweets about your mission.

As you incorporate Dark Data into your analytics, be sure to make it a policy going forward for all new data to be evaluated to ensure it is not hidden away. There is going to be more, not less, Dark Data in the future. By setting up a system to continually review your data assets, your organization will not miss out on critical insights.

One of the critical measurements of success for your organization will be how well you are able to see all your data.

The Big Data 80/20 Rule

In fundraising, the 80/20 rule means 80% of your money comes from 20% of your donors. Over the years this has become the 90/10 rule and, and with some organizations, even 95/5. With Big Data it means 80% of your data is unstructured and 20% is structured.

I would add that if we are honest, we accurately analyze 80% of the 20% of the structured data we possess. With Big Data, the goal is the analysis of 100% of the data, but it will still be 80% accurate.

One reason people cite to not move forward with Big Data adoption is they focus on the inability to ensure that it is 100% accurate. My response is we are now faced with seeing only 16% of our data accurately (20% minus 4% inaccurate data).

Big Data expands your organization's field of vision 64% (80% - the current 16%).

This reminds me of the early days of wealth screening when people would focus on who was not found rather than all the new prospects that had been uncovered. You can spend your time lamenting about the 20% blind spot or enjoy your expansive 80% view.

Even data geeks can get hung up on the missing 20%, and especially with regard to the inherent messiness of Big Data. After all, we were taught to get rid of outliers. With Big Data you are often searching for outliers (more about that later).

Traditional Data Science	Big Data Science
Dataset is complete.	Dataset is never complete.
Data quality is high.	Data quality varies from source to source.
Data is to be reported on.	Data is to be explored.
Dataset is a manageable size.	Data is big and difficult to manage.
Data Scientists focus on measuring past performance.	Data Scientists are focused on creating actionable insights.
Data Scientists omit outliers from results.	Data Scientists look for meaning for outliers.
Data Scientists see text as superfluous.	Data Scientists see text as integral to their analysis.

Understanding if your team can handle the realities of Big Data will go a long way toward helping you realize if you need to bring new members on board and/or utilize outside expertise. This does not mean your current data scientist without Big Data experience is suddenly expendable. If they have been providing good traditional analytics, then they have the foundational knowledge to learn Cognitive Analytics.

If your data scientist has been with your organization, then she has invaluable domain knowledge. If she wants to move up to Cognitive Analytics, make the investment in her education (see the chapter "So You Want to Learn More"). At the very least, she will be an invaluable colleague for your new Big Data science team member(s).

For leadership, and other insight users, you will also need to adapt to the new realities. Don't spend your time focused on what is missing or what is not perfect. Ask big questions (more on that later), and always connect insights to actions. Turning on the lights to see your Dark Data will not matter if you don't act on the insights subsequently revealed.

Big Data Management – Too Big to Succeed?

Many worthwhile projects have never gotten off the ground because they were *too big*. By that standard Big Data has little chance of garnering support unless leadership is willing to say Big Insights are worth the time and resources.

The first step for making the case to move forward is to break down the project into the primary components:

> **Data Inventory**: Identity all internal, and relevant external, data sources. Document the owners, and managers, of the data; the format (SQL; Oracle, etc.); size; frequency of updates; types of unstructured data;

Internet of Things data sources; and highlight all sensitive data.

Data Access: Identify access requirements to all structured and unstructured data from the data sources inventoried. Determine how frequently the data will be updated. Real-time may sound like the obvious answer, but reality may show real-time is, in fact, either not necessary or is too expensive to implement.

Insight Reservoir: Identify the environment where you will store, analyze, and govern your insights, as well as from which you will distribute such. On-premise, cloud-based, or hybrid is a critical decision. The staffing implications of each choice should be uppermost in your mind. Another consideration is whether you will be gathering real-time data. This will impact your choices and your costs.

Data Quality: Your data will never be perfect, but achieving the highest possible quality of data will increase confidence in the results. Certain data sources will have lower quality. Consider having a quality score for each data source, which can be used as part of your algorithms to increase or decrease the weight of a data source or data element. An example is professional data sourced from social media. This data has a higher likelihood of being out-of-date, inaccurate, or incomplete.

Master Data Management: Understanding your data across sources is essential to success. Your prospects and donors are unlikely to have a common ID across data sources, so you will need to create a "Golden Record" which tells you all of the information relating to a record. A good practice is to determine which of your

IDs will be the Master ID. As part of this process, you will also decide on the terminology, and business rules, to be used as you bring the data together. One data source might use the word *revenue* while another uses *income*. Without a common language, the analysis will be confusing to insight users.

Data Governance: We cover this in detail in a later chapter, but the key components are security; access controls; privacy policies; and applicable regulations. In a Big Data environment Data Governance has to be integrated into the technology, so activity can be monitored and reported.

Big Insights: Gather organizational questions, and evaluate them in a Big Data context. Are multiple data sources required? Is there relevant unstructured data? What are your current questions and answers? What are the gaps in insight they are delivering? All of these questions must be continually asked to ensure current realities, and future needs, are taken into account.

Actionable Insights: Mapping insights to insight users, and then to their potential actions will dramatically increase utilization. Determine which insight actions can be fully automated; automated with some insight user supervision; or only delivered to insight users for action. All of this needs to be fully documented and may involve Service Level Agreements between IT and insight users. While this may seem like overkill, it will build confidence with users, and also manage expectations.

Return on Insight (ROi): The final step is to look at how the return on Big Data will be evaluated. For financial factors establish "pre-Big Data" benchmarks.

Include the previous 3-5 year trends, so you will have a better idea of what would have happened naturally. Non-financial factors could include donor retention; alumni participation; donor satisfaction; and mission delivery metrics. Look at the questions you have been asking, and evaluate the quality of the answers as part of your ROi analysis. Another ROi will be the level of automation you have achieved because of the insights. We will explore how to calculate your ROi in a later chapter.

The technology needed to accomplish the project will depend in large part on whether you decide to implement an on-premise; cloud-based; or hybrid solution. No matter which you select, the tools involved will fall into the following areas:

Data Connection, Ingestion, and Processing: Tools to securely connect to any type of data source; ingestion of the data into a Big Data environment; and processing of the data to create a schema for analysis.

Data Administration and Security: Tools to govern, secure, and administer the transformation of data into what is needed for the creation of actionable insights based on organization, and regulatory, policies.

Data Integration: Tools for establishing relationships between datasets, including entity resolution, in order to provide a cross-organizational view of all data.

Data Quality Assurance: Tools to identify and address inaccurate, incomplete, duplicate, and erroneous data on an on-going basis.

Data Cataloging: A cross-organizational repository for all of the metadata relating to data assets. This includes all processes for integrating, securing, and governing those assets.

Master Data Management: Tools to provide a cross-organizational reference source to ensure the accuracy of data and the relevance of the insights provided.

Sensitive Data Masking: Tools to remove, protect, or obscure sensitive data, including personally identifiable information (PII) and demographically identifiable information; social security numbers; credit card numbers; health information; mission beneficiary information; and student records.

Data Security Monitoring: Tools to continually track and analyze security vulnerabilities including points of access; user activity; and accessing of sensitive data.

Cognitive Analytics: Tools to create algorithms and formulas based on the analysis of datasets. These tools need to go beyond traditional business intelligence tools, enabling such things as machine learning and cluster analytics.

Streaming Analytics: Tools to process data in real-time, or near real-time. The insights may be integrated with an alert system and/or incorporated into analysis with lower latency datasets.

Insight Reservoir: A platform to securely collect, store, manage, and analyze all your structured and unstructured data. If a data warehouse is utilized to store structured information for business intelligence and reporting, the data warehouse will become a source of data for the Insight Reservoir.

Now you have a better understanding of the key aspects of a Big Data project, and the tools you will need to be successful. Now let's tackle staffing. Depending on the size of your organization, you may have little or no internal capability to implement and manage a Big Data project. If you work within a large organization you may have some or even all of the skills currently on staff, but those colleagues may be overwhelmed with current obligations.

The following is a list of personnel you will need if you decide to fully implement and manage in-house:

- Project Manager
- System Administrator
- Network Administrator
- Database Administrator
- IT Security
- Business Intelligence Analyst
- Data Scientist
- Java Hadoop Developer
- Q&A Engineer

The cost of these positions will of course change based on where your organization is located, but also keep in mind there is fierce competition for people who have experience implementing Big Data. My recommendation, unless you have a strong IT team, is for you to outsource as much of the technical aspects as possible while keeping data science capabilities in-house as well as Data Governance management. Even if you are confident in your IT capabilities, still look at bringing in help for implementation to ensure it is done right, and minimizes the disruption of your current operations.

Transitioning your prospect research team into insight analysts is another important step in becoming an insight-driven organization. In the chapter on the changing role of prospect research and prospect development we will look at this in more detail, but the central point is to focus efforts around *actionable insights rather than data.*

You have looked at the elements of the project; the technology you will need; and the staffing. Now it is time to take an honest look at your organization's readiness to embrace Big Data.

There are many assessment tools. One I like is by Transforming Data With Intelligence™ (TDWI), a division of 1105 Media, Inc. They offer an online assessment you can take at no cost. It breaks down organizational maturity into five levels:

>**1 NASCENT** – Little or no understanding among leadership about what Big Data is or what it could do for the organization. May have implemented a data warehouse and done basic business intelligence around key performance indicators.

>**2 PRE-ADOPTION** - The organization is beginning to look at Big Data (perhaps reading this book is part of pre-adoption). Some technology may have been purchased or rented such as Hadoop to create a test environment. At this stage, the project is often driven at the departmental level.

>**3 EARLY ADOPTION** - Proof-of-concepts (POC) dominate this stage. A couple may have made it to production where insights are being acted on. A reason there can be a chasm between this stage and the next is the time it may take to realize ROI. I ran into this issue with wealth screening. While organizations were

identifying new prospects, they still had to go through the solicitation cycle. This caused an 18-36 month delay between receiving the data and realizing the benefit. The good news is once the gifts started rolling in, the adoption of wealth screening increased dramatically. I believe the same thing will happen with Big Data.

> **The Chasm** - The success of the early adoption stage will trigger a series of events, which will delay full implementation across the organization. These include the need for budget to support additional staffing and technology acquisition; political barriers, including data ownership, will need to be handled; a data governance structure will need to be created; security will need to be upgraded; and other departments who may not have been part of the POC will need to be brought into the mix.

4 CORPORATE ADOPTION - This is the major phase of Big Data adoption as the organization makes it an integral part of its operations. A cross-section of users will be using insights, and the organizational structure to support long-term adoption has been put in place. Another hallmark of this stage are changes in how decisions are made through using the insights now available.

5 MATURE/VISIONARY - Very few companies, much less NGOs, have attained this stage. Those that have are executing Big Data projects on an on-going basis and are budgeted and well-planned. Big Data has created an atmosphere of excitement and energy around the insights being derived and the potential for future insights. I see this stage as having moved from fearing the future to embracing the future.

© TDWI https://tdwi.org/pages/maturity-model/big-data-maturity-model-assessment-tool.aspx

It is interesting to note the chasm between early adoption and corporate adoption. This captures a reality of Big Data projects, which is they often begin on a small scale within a part of the organization, and then grow to a point where leadership green lights a proof-of-concept.

If the proof-of-concept (POC) takes longer than expected and/or does not produce measurable results, then into the chasm they go. The lesson is to carefully plan the POC, so the scope does not get out hand in terms of cost and time. Also, make sure the focus is on delivering an insight you know leadership cares about and the organization and staff can act on.

Bottom-line – don't start with the meaning of life. Start with something a bit more down to earth like, "How do our email and direct mail channels influence each other?"

Your ultimate goal is to have all of the parts of your organization — mission; fundraising; and operations — feeding data into the Insight Reservoir. This allows for the free flow of insight while still enabling each area to have its own processes and even unique technology to meet its needs.

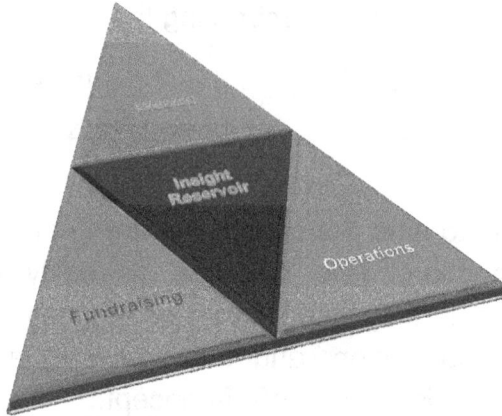

Big Questions to Go with Your Big Data

Have you ever been frustrated when you found out something you thought you should have been told? We all have. Sometimes this is because people withheld the information or thought you didn't need or want to know, but another reason could be you didn't ask the right questions.

Big Data requires real thought about the questions you need to ask. As I mentioned earlier, keep them broad with no hypotheses. Reframe a question about your fundraising from "Why are our e-mail open rates dropping" to "What is driving our email results?" This question enables you to get a wider view of what is going on. Maybe your fundraising results are doing well, and the open rates have more to do with people opening it on platforms where you are unable to get the metadata regarding action taken.

Another challenge with asking big questions is these inquiries can cause unease if not outright unrest in your organization. When you ask why something is not working, the folks managing that area are likely to get defensive. When you muse about taking the organization in a new direction, people will wonder if they are going with you.

A way around this challenge is to position your Big Questions not as critiques of the status quo or the seeking of a new direction, but rather as the basis for an on-going conversation about operations and vision. Additionally, you must continually market change to the humans we all are, by answering the following question, from the perspective of your colleagues: What's in it for me?

It is often said people are unable to be strategic because there is so much tactical work to be done every day. Your Insight Reservoir isn't worried about your day-to-day operations, so it can look at your questions unshackled by the constraints we all put on our thinking every day to get things done. Your Insight Reservoir is not worried about what's in it for its own benefit.

How do you discover your big questions?

You might think you call a meeting or form a committee, but I would start with soliciting ideas from everyone in your organization. You might be surprised what lower-level personnel see from their perspective, and it will help create buy-in.

Leadership could send an e-mail asking, "If you could ask three questions about our organization what would they be?" This could be followed it with "If you could ask three questions about your department what would they be?" You may get better results if you allow people to send their responses anonymously. You could also use a "graffiti wall" in your organization, in a secure location of your office – large paper posted in the hallways asking these same questions can prompt your colleagues to share.

Now bring your leadership team together and use the questions to get the conversation going. If all goes well some of the more uncomfortable questions will not have to be put forward by the team, and once they see it is okay to think differently they will join in.

Organizations exist in a dynamic environment, so keep the door open for questions. Create a virtual "Question Box."

> **Small Data Question:** How much did we raise last year?

> **Big Data Question**: Why did we raise more (or less) money last year?

> **Small Data Question**: How many people liked our school's Facebook page?

> **Big Data Question**: Who is wearing our merchandise in their social media pictures?

The quality of your questions coming in will go up as people see the value of the insights coming out. Share results as widely as possible. By clearly connecting insights to actions, and actions to results, you will become an insight-driven organization.

I was working with a private day school, and they had become concerned their marketing program had seen a fifty-percent decline in lead generation. Leadership wanted to know what was wrong with marketing. Were they not doing enough outreach? Was the copy not resonating?

This was in the middle of the Great Recession, so I looked first at the economic conditions in the area the school served. It turned out the number of families who could afford the school had declined by fifty-percent, so their marketing program was performing as it always had with their constituency. The problem was their constituency had lost half its members.

The Big Data question leadership could have asked is, "What are the drivers of the number of people interested in our school?"

Embracing Big Data questions does not mean letting go of the questions you are currently asking your small data systems. You will still need to know all of the basic quantitative answers, but now you can understand what the qualitative drivers are for those numbers.

04

Cognitive Computing – The Brain That Turns Big Data into Insights

"A computer would deserve to be called intelligent if it could deceive a human into believing that it was human."

—Alan Turing

Big Data is the smokescreen obscuring the real technological breakthrough of this century – Cognitive Computing. Think of Big Data as Big Storage – without it, cognitive computers could not work, but without Cognitive Computing there would not be much point in storing all that data.

Cognitive Computing came of age in 2011 by playing and winning a game. IBM Watson achieved what Alan Turing had imagined – the ability of a machine to interact with the world as if it was human (in this case *Jeopardy!* champion Ken Jennings). Being able to understand natural language (the question); store and retrieve the best possible answer (Wikipedia and Encarta); and then know if it should answer the question (or, rather, provide the answer as a question) given how it was doing in the game at that time.

The last part explains why Watson gave a few responses that made people wonder if it was really all that smart. It was using a low confidence answer because it was behind, exactly what a person would do in the game.

The model for Watson was our brain. An interesting point is we don't fully understand how our brain works. When we created machines to replicate what our muscles did, we knew everything about how we accomplished a physical activity like painting a car or putting a cap on a bottle. The field of cognitive science is attempting to bridge this knowledge gap by combining the study of the brain with computer science. This is where you will find the frontier of Cognitive Computing.

What Watson did was replicate what we do know about how our brains operate, by enabling the right-brained processing of language, images, and concepts to work with the logical, linear, and analytical thinking of the left brain. We will go into this in more detail in a later chapter, but here is a way to look at Cognitive Computing as we look at our brains:

> Big Data is where all the data is stored, managed, and governed; analytics is the left side of the brain performing the calculations; natural language processing is the right side of the brain, turning all the messy unstructured data into structured information so it can be analyzed by the left side; machine learning is where the thinking happens, constantly creating new insights, and challenging old ones; deep learning takes

this a step further where unsupervised learning occurs based on multiple layers of insights created by machine learning; modeling and processing is where the insights are joined to your organization's outcomes; and finally the applications are where all of this is visualized, so you can actually see your insights.

There are three hallmarks of Cognitive Computing:

1. Understands natural language as spoken by humans.
2. Is able to generate and evaluate an evidence-based hypothesis.
3. Adapts and learns based on training, interactions, and outcomes.

The theoretical work on this was done decades ago, and I have had the privilege to meet some of its pioneers. They had tried to create an IBM Watson many times, but the storage, processing, and visualization tools of their day were not up to the task.

It was not until Hadoop came along, with its ability to store any type of data efficiently, and processors exponentially increased in power, that the theories could become reality.

A final piece of the puzzle was storage becoming cheap. Today, thanks to Amazon Web Services®, and other cloud providers, you can rent terabytes of storage for what it used to cost to own your own 500 Gigabyte server. In fact, according to the Statistics Brain Research Institute, in 1980 the average cost of a GB was over $400,000 while in 2016 it was only $0.019.

Cognitive Computing has become the platform on which everything else operates. It is the mind of computing going forward, able to orchestrate all of the capabilities to produce insights much as you are doing right now as you read this book.

To comprehend the meaning of the future it is a good idea to understand the difference between it and the past:

Relational Database	Big Data Lake
Master-slave architecture	Distributed architecture
Low to moderate data velocity	High data velocity
Data ingested from one to few sources	Data ingested from many sources
Optimized for structured data	Optimized for structred, semi-structured and unstructured data
Need to manage data volume by purging/archiving	Retain data as needed
Schema created pre-data load	Schema created on data load
Data relationships established by users	Data relationships discovered by database
Uptime managed with failovers	Distributed architecture protects uptime
Single datacenter or cloud region	Multi-datacenter and multi-cloud environments
Keyword searching of text	Natural language queries of both structured and unstructured data
Data mapping required to create datamart/warehouse	No data mapping required
Designed to support traditional analytics	Supports maching learning and deep learning

You can see just how archaic the databases we have been (and are still) using are. No wonder they have struggled to deliver value to end-users. The men and women of IT have been the MacGyvers of their organizations, finding ways to deliver what was needed often just in time or at times the MacGrubers (of *SNL* fame) delivering it a little too late.

Big Data has its Five Vs, so I propose Four Cs for Cognitive Computing to complement them:

Curiosity: We need to be as curious as a child, and as uninhibited. We also need to fight the urge to just ask the same old questions, hoping for new answers. Cognitive Computing is not about a new chart for your Business Intelligence dashboard. Only with Big Questions will we discover the Big Insights.

Consequences: Focus your cognitive projects on areas where the return on investment justifies the resources expended. Connect them to the people and processes able to act on the insights gleaned. And most importantly, make people accountable which means having both positive and negative consequences.

Collaboration: Cognitive Computing is a team sport. Bring together subject matter experts; data scientists; IT; and the people who can act on the insights. Getting this diverse group to successfully work together will determine the success or failure of your cognitive projects. As you build your cognitive capabilities, the final collaborator will be the cognitive applications themselves.

Clarity: Being able to present insights to end-users in clear, non-technical, terms is essential. Break away from traditional charts, graphs, and reports. Think of yourself as a storyteller. When using graphics, think graphic novel, not a pie chart. Cognitive Computing is enabling a deep understanding of all the data available. We will need clarity if we are not to get lost in its wonder.

Now let's look at different aspects of Cognitive Computing to understand what each can do for your organization.

The Left & Right Side of Data

The technology we depended on until Big Data and Cognitive Computing came along was perfected to serve the left side of brains. That is where our logical analytical thinking resides. On the right is where our passions live. The good and bad angels on our shoulders are in fact the left and right sides of our brain. The left is constantly pointing out the logical reasons to do or not to do something. At the same time, the right is making an impassioned case for why you should throw logic and reason to the wind, and follow your passions and interests.

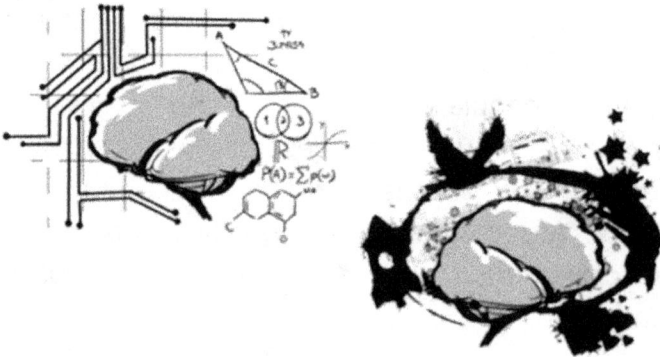

©2017 Max Lawson

Studies have clearly shown philanthropy's home is on the right side. When you help others, the same parts of the brain light up as they do when having sex, taking drugs, and listening to your favorite music. This means philanthropy is in an unstable place where decisions are made not based on numbers and facts, but rather feelings and passions.

Dashboards were a way to bridge the gap between the logical left-brain and the passionate right-brain. The thought was if we just create pretty pictures of all those numbers, then the right-brain end-users will see what the data is telling them. Unfortunately, this was not the case.

The passionate right-brainers in the room would point out why the numbers were wrong, and they would do it by using qualitative data. This drives analytical left-brainers crazy because it is often subjective or *soft* rather than the hard numbers. What we all need to accept is philanthropy is driven by forces that are often subjective.

Giving, especially at higher levels, is not a rational decision. People support your organization because they believe in your mission, and want (really, hope) your organization to succeed. Donors might involve their rational left-brain accountant, but the ultimate decision will be based on the siren sounds coming from their right-side telling them how good they will feel by believing in others and seeing the impact of their own generosity.

Philanthropists are motivated by helping the promising student who drops out of school in their junior year because they don't have the funds to finish or the person who loses their job not because they were not qualified, but because they are addicted to alcohol. On a traditional dashboard, these show up as lower retention rates and unemployment. Cognitive Computing enables you to bring qualitative data into the story, enabling you to understand what is driving these results. These drivers are often hidden in the notes of school counselors and caseworkers.

Later we will discuss Cluster Analytics, which place the troubled student and the job trainee into groups sharing their characteristics. This has two positive outcomes: first, you are able to help groups which may now be hidden in the numbers; and second, you are able to look at your overall numbers in the proper context, keeping you from drawing the wrong conclusions.

All the talk of left and right can make you think of our brains as a tennis match where we are constantly sending information, thoughts, images, concepts, and ideas back and forth. How is all of this input managed without driving us crazy?

Enter the largest part of our brain, the corpus callosum. This is where the rational left side and irrational right communicate with each other, enabling us to come to conclusions, make decisions, and to act on them. Cognitive Computing's goal is to be able to do all of these things. It is not there today.

My belief is we need to maintain the corpus callosum role even when Cognitive Computing becomes fully functional. It is not that we won't let our Cognitive Assistant make decisions on its own – such assistants already load Walmart trucks and decide what we see on Amazon and Netflix® - but because of the iterative nature of the technology, and the realities of human and organizational processes, we need to keep our own cognitive computers (brains!) in the mix.

Understand your organization from the brain's perspective - separate aspects of your mission and operations by whether they are left-brain or right-brain elements. Amount of gifts is a left-brain item; a thank you letter for a gift received is the right-brain. The number of meals you serve at the shelter is left and the reviews of the soup are right. This exercise will also help you see where right-brain – qualitative — data is needed to help you better understand what is driving your left-brain results.

You can also begin to measure the quality of your data during this process. As you look at sample data, especially of the right-brain information, you may discover some of it lacking and/or not well-connected to the left. Call reports are a good example. They contain (or at least should) the answer to *how* a meeting went; *why* something did or did not happen; *what* they are truly passionate about; and other insights not found in the left brain outcomes such as how much was raised.

What you will find is as your organization provides the benefits of Big Data, the quality of data will increase. It is the natural outgrowth of people seeing their data turned into insights.

Teaching not Programming

Think of Cognitive Computing as a really bright child. There is tremendous promise, but her full potential will never be fully realized without an education.

With a child, you push, prod, and cajole them towards a particular future: doctor; lawyer; artist; musician; or an athlete (or all?). Without the Williams sisters' father constantly teaching and motivating them, they may well have still been good tennis players because of natural talent, but it is doubtful they would have been great ones. Unlike a child, Cognitive Computing doesn't start with an aptitude for any one career. This means the possibilities are limitless, but it still requires choosing a path and a good education.

Watson won *Jeopardy!,* but it still had to be taught how to read an X-ray. In fact, there is no *one* Watson; there are as many as there are challenges to meet. A common misperception is Watson knows all, when in fact an expert in any one category of *Jeopardy!* would likely beat Watson unless Watson was trained even deeper on that subject.

Your cognitive platform will learn as a child does, going through stages of cognition. It will try your patience like a young child, constantly asking questions and then asking why when you give answers. You will have moments when you are genuinely concerned your little "Cog" is not as smart as you thought. Even simple questions will stump it. Then seemingly overnight it will know the answers exactly as a child struggles with multiplication tables, but then something clicks and they are off to division.

Cognitive technology, like your brain, is constantly working. That guitar lesson which seemed to go so badly yesterday has resulted in you playing a song today you couldn't before. That is because your brain was still doing the chord changes and strum patterns as you slept. With us this is called *deep sleep* - with little Cog it is *deep learning.*

How do you teach a machine? First, you are not teaching the machine to know everything. You are teaching it about a specific domain or area of expertise. Just like you wouldn't want a lawyer performing your heart surgery, you don't want a cognitive platform built to win *Jeopardy!* deciding if your driverless car should stop for the person crossing the street in front of you.

Second, you must provide enough data to enable the machine to understand both what the data is, and what it is not. The good news is you don't have to provide every answer to every question, but you do have to provide enough to allow the machine to learn on its own. Think of dog breeds. You provide 1,000 pictures of Labradors and 1,000 pictures of dogs that are not Labradors and the machine is then able to look at pictures of dogs and tell you if it is a Labrador or not with a high degree of certainty. Provide more data and the certainty increases.

You may be concerned your organization is not smart enough to teach a cognitive platform. Don't worry — as long as you know about your organization, and the domain within which you operate, you are fine. In the chapter on Natural Language Processing we will discuss ontologies, and how they are the key to unlocking the value in unstructured data. For now, think of an ontology as the Rosetta stone informing your cognitive platform what data means to you.

A good way to get comfortable with teaching a computer is to think about how you learned about your organization; your sector; and your function within the organization. Find material relating to each aspect, and this will form the corpus. You visited your organization's website; gained information from conferences and workshops; you may have gone through training from HR; and you read trade journals.

All of these elements are the foundation of a training corpus. You will use it to build a curriculum for your cognitive platform. Remember, the subject you need to be an expert on is not brain surgery (unless that is your organization's focus). It may be education or disaster relief. It is something you, in conjunction with your colleagues, know.

Confirmation Bias - What's yours?

Bias is a loaded word. When someone says you are biased your knee-jerk reaction is to emphatically say no! You talk about how open-minded you are about people, places, and pretty much everything.

The reality is we are all biased. The reason is simple – we could not survive if we did not have biases. Imagine waking up in the morning and wanting everything for breakfast. Your eggs any way; toast, bagel, biscuit, bring them all on; and cereal - you like granola just as much as you like Count Chocula™. Now think about liking every style of furniture, and every color exactly the same.

You couldn't function if this was your reality. This is one of the reasons we enjoy having favorite foods, styles, and colors. What we don't so readily admit is those same biases are brought to decisions we make in business. This explains why in a meeting when someone brings up something truly new there is a big pushback with people defending the status quo and/or pointing out all the reasonable and unreasonable reasons the idea will never work (or at least not at your organization).

Enter artificial intelligence (AI), perhaps the best-known element of Cognitive Computing. It can conjure images of crazed robots taking over the planet as well a sweet voice telling us to avoid a hazard up ahead on the highway.

One of the most powerful aspects of AI is the technology does not bring our ancient bias instinct to its work. To AI, the status quo holds no sentimental value, and a new idea no sinister motives. It is true AI can be taught to have a bias, and all data it consumes has at least some bias to it. In fact, teaching AI to understand your domain or organization is essential to making it valuable.

What AI will not acquire is a bias towards one piece of data or insight over another. It will not play office politics or seek to enhance its own agenda over others for its own reward. This can make for uncomfortable moments as your cognitive platform presents an insight fundamentally altering the way you do business.

To fully assimilate this technology we will need to become comfortable with uncomfortable insights and facts. We also need to ensure that those insights are only made operational once your organization is ready.

It may be a fact you need to hire or fire personnel or stop doing some tasks, and it may also be a fact you are unable to make the changes immediately. Think of your cognitive platform just as AT&T® saw Bell Labs. It is a place where you are constantly challenging the status quo, but you control when, and which, insights are acted upon.

I put this forward at a conference not too long ago and had the good fortune to have someone from the current owner of Bell Labs (Nokia®) in the room. She said I had captured the spirit of how it is used by the corporation. Until now the idea of every organization creating its own Bell Labs was unthinkable. Cognitive Computing has made it possible.

Pull together your team and ask these questions:

- What is your favorite kind of food?
- What is a kind of food you will never try?
- What is a food you think you might like, but have never tried?

With everyone relaxed, ask:

- What is your favorite part of your job?
- What is a job you would never do?
- What is a job you might want to do, but have never tried?

Then ask:

- What do you most like about our organization?
- What is something you believe our organization will never do?
- What is something we do today that you would fight hard to make sure we keep doing?

On the food questions, my answers are short ribs; any kind of stinky cheese unless I absolutely have to or I'm tricked by my wife; and ramen noodles in Japan (I am no fan of the U.S. version).

By starting out with something non-threatening like food, you can introduce the concepts of bias without getting everyone defensive although I will warn you — I take a lot of heat for not liking stinky cheese. You might challenge yourself and the team to try something they believe they do not like and have never tried at your next team dinner.

The goal, of course, is to get to the biases holding your organization back. It can be the feeling your organization never tries anything new or has always done things a certain way. You can compare this to always going to the same restaurant and always ordering the same thing.

Helping people appreciate the role of confirmation bias in their own decision-making will enable them to move in new directions. The fear they feel is real, and needs to be acknowledged. When we are faced with doing something we have not done before it triggers ancient responses that protected us from being killed by a saber tooth tiger.

Your new idea might not have sharp claws and teeth, but it does represent something unfamiliar. Cognitive Computing at its best can help us overcome our fears by venturing out of our data caves and providing us a safe path to a new destination.

The ROi(nsight) of Cognitive Computing

The value of data is directly proportional to its ability to provide actionable insights, and those insights are acted on. Take any of the three elements (data, insights, action) away, and you are unable to realize a return on your investment in Cognitive Computing.

We have come around to the idea of data being an asset rather than a cost, yet for the asset to have value it must be connected to actions. All projects need to start with the actions and work back to the data.

- Increase the number of donors giving more than $1,000 per year.
- Increase the number of members who become donors.
- Decrease the donor attrition rate.
- Increase the number of people helped by our mission.

Armed with your goals, you turn to your data. This is where the Insight Reservoir comes into play. By having all of your internal and external data in one place, you cast the widest possible net for potential points of correlation. This is key because unlike with traditional analytics where you start with a hypothesis such as, "past giving determines future giving," you make no assumptions about the answer. This will create your actionable insights.

Next, you connect insights to the action(s) needed to realize its value. This may require the actions of multiple people within your organization. Connecting the insights to the people who can act on them is critical to success. The final step is measuring the actions, so you can see if the insights or the people acting on them are causing positive or negative results.

An example from my career is wealth screening. When I started P!N, I thought for sure every fundraiser would be thrilled to receive millions (and sometimes billions) of dollars in new potential. It turned out making those cold (or even lukewarm) calls was not something for which every gift officer was waiting breathlessly.

The lack of action on wealth screening results almost ended my first business. It was not until I started a workshop (Wall Street for Fundraisers) on using the data that people really started to embrace it. And of course at some point a few intrepid clients raised millions, and then everyone wanted to have their file screened.

This same pattern will happen with cognitive projects. No matter how compelling the insights, they will more than likely require a person to do something. My mother's story of the four frogs on a log has guided me – if one *decides* to jump, there are still four left. It is only when a frog *actually jumps* that anything changes.

So, you have decided to identify your most important goals, gathered all your internal and relevant external data, set-up your machine learning environment, and are ready to connect the insights to the people who can act on them. What is the return on all this effort going to be?

Let's start with how to value data. At its most basic level, a data element such as a constituent's record can be valued by adding the cost of replacing it and the economic value to your organization. You could take your acquisition cost plus your average gift to establish your Basic Data Value.

If you want to get a more accurate valuation, factor in the quality and exclusiveness of the data; its value to your organization; and any market value (list rental for example). This is not about getting a value to the penny; it is to enable you to see data as an asset rather than just a cost.

So far we have been working with traditional structured data found in traditional environments like a donor database. You can do this valuation now, and you should even though this is only level one of the data valuation process.

There are four more levels of value, and none of them can be reached without Cognitive Computing:

> **Level One**: Data managed in individual databases with little or no connection to data managed in other databases.

> **Level Two**: Data transformed into Information by bringing multiple datasets together in a Data Lake (a data warehouse does this, but only with structured data).

> **Level Three**: Information transformed into Knowledge by establishing connections within the datasets to create a Golden Record with all available information on any particular constituent available as a single record.

> **Level Four**: Knowledge transformed into Insight by using cognitive analytics to find correlations between the datasets connected to organization actions.

Level Five: Insight transformed into Wisdom by acting on the Insights to derive value for your organization.

Now we are ready to calculate our Return on Insight (we will use ROi to distinguish it from the traditional ROI). The ROi puts the value on the *outcome* rather than the data and insights themselves.

Let's say your organization's average gift from direct mail is $50, and your response rate for renewal mail is 4%. This means for every 10,000 bad addresses, you are losing $20,000 (400 gifts x $50). Now if you really wanted to understand the cost of a bad address you can factor in the average number of years a donor gives to you, so your formula might be ($50 x 400) * 3 = $60,000.

Let's say it cost $0.025 per record to run a National Change of Address, and you ran it every year for 3 years. That means it will cost you $750 to run NCOA ($250 per year), giving you a $52,500 gross profit. Even the hardest hearted accountant will see NCOA not as a cost, but rather as essential to the health of your organization.

If you have a good planned giving program, then consider incorporating the percentage of donors who make a planned gift, and the average value of those gifts. And don't forget major gifts. How many of your donors go on to higher levels of giving, and what percentage of your fundraising is coming from those donors?

Given this math, can you believe anyone argues about spending money on having good addresses? It has always baffled me, especially when you consider most organizations actually lose money to acquire a donor. This means without renewals direct mail makes no sense at all other than for companies charging for the mail piece creation, printing, mailing, and collection of the funds.

How might Cognitive Computing help you with address changes? The most obvious is to ensure you are always on top of any return mail, perhaps even sending via an Application Programming Interface (API) the bad address as soon as it is received. A less obvious one is to study the history of people whose addresses change to see if there are any changes in other aspects of their life or data within your system which could give you a clue an address change has happened or is imminent. Perhaps you see a job change to a company in another location or you see giving in a new city. You might actually have been told they had moved or are planning to move in a contact report, but that information has not been changed in your Constituent Relationship Management (CRM) system.

Going through the process of establishing a measurement for ROi will ensure everyone in your organization understands how essential each person is to realizing the value. Too often data, and insights, are treated as if they can deliver value on their own. I used to say wealth screening delivered unsigned checks. It was up to the gift officers to secure the commitment.

Depending on the insight being valued, there may be multiple dimensions. It is said people who like your organization will tell 3 people; people who don't like your organization will tell 5; and people who don't like your organization, but you work to repair the relationship, will tell 10. This gives you an interesting X factor to increase the value (or the damage) of a happy or unhappy constituent. In the case of social media, you can use this to increase the value of people who have large networks and are spreading your story.

For non-data folks, all these formulas may not seem very relevant to getting anything done. Keep in mind without actionable insights you may find yourself underperforming not because of lack of effort, but because *you are doing a great job on the wrong things.*

The ideal group to build your ROi formula is a combination of your insight team, action personnel, and leadership. Work together to come up with the actions which will yield the most important results; the data you will need; and identify the people who will act on the insights.

When you come up with your formula there will surely be people who will say, "We will never realize the full potential," and/or, "What if these assumptions are wrong." It is confounding how much we want certainty when we all know nothing except death and taxes are a 100% bet.

If people get stuck on the need for perfection, use an old rating trick – a range. This shows everyone the figure is an estimate. For the address correction, you could have used $30,000 - $60,000. Both are still a good bet against a $750 investment.

Once an ROi formula has received agreement, create a simple spreadsheet and make it available to everyone who needs it. When you embark on a project, have the numbers right next to your other goals. Better yet, why not just have your Insight Reservoir do the calculation for you.

When the numbers come in much higher or lower than expected, don't immediately adjust the formula. Look deeper at what drove the changes. Perhaps actions were not done at the level you anticipated or there was an economic event, which impacted all of your efforts. If you are unable to find a reason for the differences, then you can adjust the formula up or down.

An organization's ability to blend art and science has never been more crucial to its success. Determining ROi is one of the ways to infuse this into your organization's culture.

CHAPTER

05

Natural Language Processing – Using Your Words

> "To handle a language skillfully is to practice a kind of evocative sorcery."
>
> —Charles Baudelaire

All the talk about qualitative data analysis is for naught if you can't understand language as it is spoken. That is what Natural Language Processing (NLP) is all about. Just as you know what a person is saying when you hear, "I'm hungry, I want an apple" vs. "I really want an Apple instead of a PC," so now can a computer.

There are two parts of NLP: Natural Language Understanding (NLU) and Natural Language Generation (NLG). NLU breaks

73

down text into its parts, the same way your brain does. It recognizes nouns, verbs, and adjectives as well as entity types: people, places, and things. NLG is then able to generate natural language based on a knowledge base.

In practical terms, NLU enables text mining of all your notes, white papers, reports, social media and proposals without you having to categorize, label, or tag the information. NLP brings all of this qualitative data alive and enables it to be processed with traditional structured data. NLG is then able to create copy for email, marketing copy, and any other communication where it has a knowledgebase from which to work.

How is this done? Let's start with NLU.

Understanding parts of speech is relatively easy — knowing Bob Smith is a person's name, and Microsoft is a company, is not difficult. What is much harder is knowing Bob Smith is a professor at your school or the Mary Smith Center is the name of your performing arts center. For this to work, you will need to build an ontology.

You may be more familiar with a taxonomy, which classifies terms in a field. The National Taxonomy for Exempt Entities is the best known in our world. An ontology takes the taxonomy into the realm of human speech where there can be more than one way to refer to the same thing. Let's look at the term *climate change*. There is also *global warming*. You might want both terms to be under *environment*.

Wine is a good way to understand how all this works. When thinking of a bottle of red wine at the highest level it is simply Wine. Next, it is Red Wine – then there is the varietal – cabernet franc; then a Region – Napa Valley; then a vineyard

– Peju®; and then a year. Once you have an ontology, NLU can understand, "I really enjoyed the 2012 cab franc from Peju." Notice they didn't say Cabernet. That works because the ontology includes different ways a word is used.

At this point, you might be concerned it would take you too long to create a complete ontology. The good news is you don't have to. You can focus on just the people, places, and things that are unique to your organization. NLU has built-in libraries it can use to identify everything else.

Here is a potential ontology for a Giving Interest:

1) Giving Interest
   a) Environment
      i) Conservation
         (1) Land
         (2) Ocean
         (3) Rivers
         (4) Lakes

The ontology will define the entity types within the text: Person; Country; Location; and Organization are examples of entity types. NLU applications will have most of the common entity types, but your ontology will enable a much deeper look at your text. A museum might add Collection, Artist, and Owner to entity types.

Sentence diagramming, just like you learned in school (or not), is a foundational element of understanding language. Is it a noun, pronoun, adjective, adverb, or verb? NLU is also able to separate a sentence into Subject, Object, and Action:

Bob and Mary   want   to donate to the science center

Subject   **Action**   **Object**

We take this ability for granted, but for NLP to be able to both recognize a word and how it is being used in context is why we are seeing incredible breakthroughs in the application of NLU to real-world problems.

Another potential obstacle is what to do with foreign languages. Here again, the technology is moving quickly, and today many languages are available. You will need to have ontologies for the languages, so that your unique dictionary is applied to all of them. That sounds complicated, but it is not. You just run your ontology through translation software, and then have a human edit it. That way you are not creating each one from scratch.

The payoff for your efforts could be huge if you have an international constituency and/or have a lot of relevant data in foreign countries.

As with all cognitive technology, it will get smarter over time. As it gains experience with your data, and you teach it, the results will be become better and better.

No matter how good your NLU is, without text to analyze, it will not be able to accomplish much. Here are some potential sources of unstructured text:

- Social Media
- Class Notes
- Call Center Reports
- Prospect Call Reports
- Note Fields in Relational Databases
- Research Profiles
- Field Reports
- Organization and Sector White Papers
- Websites
- Mission Reports
- Organization and Sector Publications

Once you have your NLU program humming, it will encourage you to think of ways to obtain more text. One easy source is an online survey with a couple of engaging open-ended questions.

For schools, you can take full advantage of the strong feelings reunions can engender. For years there have been class notes, and those can be interesting (I find it a little disconcerting that my year keeps getting further and further back in the school's magazine), but without the right prompt, they tend to be facts rather than feelings. Companies like Meno® are addressing this by creating digital reunion books where people can write short stories about their lives, and what the school meant to them.

With NLU, this type of rich content is no longer just something nice to have. It provides actionable insights about individual constituents and can inform the larger conversation about what is important to your constituency as a whole. You can be sure if one person loved a professor, a lot of others do too.

The University of Michigan is using NLU to help students develop their writing skills:

> After students write and submit their essays, automated text analysis (ATA) is used to evaluate the essay, looking for the qualities of good writing that have been built into the algorithm. These qualities are examined using a variety of text analysis techniques, such as vocabulary matching or topic matching, which the algorithm detects. ATA generates a predicted score, which is sent to ECoach for a writing fellow to verify. After this verification, the score will be made available to students. https://worldclass.umich.edu/m-write-to-include-automated-text-analysis-in-upcoming-fall-semester/

Note how writing fellows are verifying the results. This is an example of augmented, rather than artificial, intelligence (more about this later). In many applications of Cognitive Computing, you are going to want a human in the mix. A student receiving a low score on her essay without also receiving a plan of support for helping to improve could increase rather than decrease dropout rates.

You are most likely not a student, but wouldn't you like to have an application like this to improve your writing? Checking spelling and grammar has been around for years, but having real-time feedback on the quality of our communications is what we will soon come to expect.

For fundraising, a potentially huge benefit of NLU is it will encourage much better call reports from frontline fundraisers. Over the years it has been a running joke that call reports are nothing more than a few words to make sure you get your expenses paid, and activities counted. Here are a few gems:

"Good visit"

"Nice house, new car"

"Has three dogs, and a mean cat"

This last one is not very useful unless your NGO happens to be a humane society. Once your team knows you can actually mine their reports, you can ask for details about interests, passions, and what is really motivating their giving or non-giving as well as doors they can open for you.

Now we turn our attention to the other side of NLP – Natural Language Generation (NLG).

The generation of language based on a database of language is not a new concept. What is new is the exponential increase in the amount of language stored as data, and the organization of the data using NLU.

We are starting to see adoption of NLG at the enterprise level, and this has wide implications for our sector. One of the first applications of NLG for philanthropy was developed by a start-up, Gravvty™. Using the data in your CRM, and examples of your emails, their app (appropriately named, First Draft™) writes the first draft of an email to a constituent. This is just the beginning of what will be available as NLG begins to be adopted across all aspects of organizations.

As with everything in Cognitive Computing, there are levels of NLG:

- **Level One - Merge**: The mail-merge function in your word processing software is an example of the most basic NLG where you are taking data such as a person's name and inserting into a letter at a specific point.

- **Level Two - Template**: Using templates, business rules, and very basic calculations NLG is able to automate the creation of content with more variability than Level One, but lacks any understanding of the domain or writer.

- **Level Three - Context & Conversation**: This is where NLG is able to write contextually about a subject drawing on data from multiple sources, and the writing is in the conversational style of the writer.

CRM companies, with their wealth of information about constituents, are among the first to take advantage of Level Three. Now imagine tapping into your Insight Reservoir. You would have all the data from your CRM, and all the information from both internal and external sources to create rich NLG communications.

NLG is also enabling powerful chatbots to be deployed. People are going to be more willing to let go of human help-desks if the interactions are more relevant, conversational, and, yes, helpful.

Think about having the ability to quickly generate impact reports on any aspect of your mission. Constituents could just request the reports on your website, creating self-service stewardship. How would instantly-available information regarding the impact of their donation influence donor satisfaction and retention?

People in the philanthropic community have never been at a loss for words, whether it is telling the stories of those in need or describing the impact of supporting an organization. With

NLP we now have a way to turn all this rich qualitative data into deep insights and communicate those insights at scale.

Sentiment Analysis

We have had the ability to store words in a database for decades. Reading them, we are able to infer whether a person is happy, sad, upset, or indifferent. Now technology can do the same with sentiment analysis.

Early versions were based on keywords. This proved problematic because words can have different meanings, and without understanding context can be very misleading. Today, sentiment analysis is able to discern sarcasm and be taught how to look at a word from the context of the organization rather than from a general perspective.

An example is the word *cancer*. There is nothing positive about cancer unless you put *survived* in front of it. Also, the American Cancer Society might not want to see the word cancer as negative in any circumstance because it is so often referenced by their constituency. They will put more emphasis on what is said around the word.

> "I survived cancer."

> "I so appreciate what the American Cancer Society did for my husband who passed away two years ago after a long battle with prostate cancer."

Gauging the sentiment of your constituency is not a new idea. Paper, on-line, and telephone surveys have been used for years to assess how happy (or unhappy) people are with your organization.

The problem has been these surveys mostly consist of questions like "on a scale of 1 -10..." or "which of the following best describes your feelings about..." and provide possible answers from which to select. This gave some level of insight, but did not allow the respondent to give context to their answer.

You might like a restaurant's food, but hate the service. You might give them a 5 on a scale of 1 – 10. The restaurant would not know about the rude waiter who ruined an otherwise wonderful meal. In your organization, this could manifest itself as people who like your mission, but who are frustrated because your organization sends them too many solicitations.

The ability to analyze natural language encourages the best practice of asking open-ended questions. The restaurant can ask, "What do you like the most about our restaurant?" and, "What did you like the least?" Your organization can ask, "What is the most important part of our mission to you?" and, "What do you like (or dislike) about our fundraising?"

The best sentiment analysis will provide the degree of sentiment rather than just Positive, Negative or Neutral. This enables your analytics to be more nuanced, yielding more accurate insights.

The scale will also help you see sentiment over time, enabling you to understand if your organization is moving in the right direction. It is impossible to be perfect, so what you want is a clear picture of your overall performance. This will help you prioritize new initiatives, and see where current initiatives are moving the needle in your preferred direction.

Later we will explore Journey Analytics, where sentiment can be used as part of your evaluation of constituent touch points, and Cluster Analytics where you can see the sentiment for different segments of your constituency.

Striking the Right Tone

We have all read an email and inferred a positive or negative tone. We have also sent emails and wondered if we had given the wrong impression. You may have formed an opinion about the author, and your response reflected it or received an unexpected response because the reader misinterpreted the meaning of your message.

If we are honest, our accuracy divining tone from written communication is far from perfect. We are much better when we can hear the person speaking. The earnestness of a statement, or the sarcasm, comes through clearly. I have found myself using "!," and even the overused smiley/sad faces to ensure people know my feelings about a subject.

Enter Tone Analysis, a technology to discover a lot about the people who are communicating with you, and about the communications you are sending. The tone is broken down into three parts plus a special analysis for conversations:

> **Emotions**: Joy; sadness; anger; fear; and disgust.

> **Language**: Confident; tentative; and analytical.

> **Social Tendencies**: Openness; agreeableness; conscientiousness; extraversion; and emotional range.

> **Conversation Tone**: Frustrated; sad; impolite; neutral; polite; excited; and satisfied.

This combines linguistics with the psychology of language. This is another example of how Cognitive Computing has been taught rather than programmed, and the syllabus for that teaching is the same used at colleges and universities.

I promised not to dive too deeply into any one aspect of Cognitive Computing, but it is important to have a basic understanding of how this works. Emotions and language are derived by looking first at individual words: amusement or delight = Joy; iffy or undecided = Tentative. Next, you look at how words are grouped into phrases: The poor lighting made me concerned for my safety = Fear; I believe in myself = Confident.

Social tendencies are based on the Big 5 personality characteristics (extraversion, agreeableness, openness, conscientiousness and neuroticism) developed by psychologists (Costa & McCrae, 1992, and Norman, 1963). Words and phrases are analyzed to determine how much of your personality fits each of the Big 5. This means there are actually many more than 5 possibilities. You may not be open or are an introvert. You also may be conscientiousness sometimes, but not all the time.

NLP results will be presented on a scale such as:

> < .5 = not likely present
>
> > .5 = likely present
>
> > .75 = very likely present

Note: This is from the Watson Tone Analyzer.

Tone Analysis will make its best determination based on the information you have provided. You may have enough information to accurately predict one emotion, but not another. When doing your analytics be sure to factor in the confidence score, and even create business rules to omit data when the score is below a certain threshold.

Here are some ways you could put tone analysis to work for your organization:

Donor Personas: Layer in tone to personas to provide dimensions such as *happy*, *angry*, and *disgusted*. Now you can segment not just by demographics and giving behaviors. Imagine having a list of "Angry donors who gave you $1,000 or more in the last year."

Direct Marketing: Create the copy, and images, to match personas. Having a clearer picture of the personality of the person with whom you are communicating enables you to tailor your copy to meet them where they are instead of where you wish they were.

Intelligent Listening: Mine all of your unstructured data including social media; verbatim survey responses; emails; and reviews to gain insight into what people are feeling about your organization. This will inform every aspect of your operations, from the person who answers your phones to what you are posting on your Facebook page.

Call Centers: Measure constituent satisfaction, problems, and concerns. Understand these metrics as they relate to specific aspects of your organization. Assess call center agent performance in the context of the type of calls being handled.

No matter how good you are at discerning a person's tone, you can't scale your abilities (or your staff's) to keep up with your constituency. The cost of being tone deaf is too great to not take advantage of this capability. Yes, it will not be right all the time, but accept none of us humans are right all the time either.

Speech Recognition – Turning Audio into Text

Audio analytics has become much better as Speech Recognition technology has matured. You can thank Apple's Siri™ and Amazon's Echo™ as well as Dragon™ software from Nuance. To make these applications work, sophisticated algorithms had to be built to understand human language. Once the accuracy was at 80+%, audio analytics was born.

Accurate Speech Recognition has opened up all of the NLP tools from Tone Analysis to Personality Insights. Not surprisingly, call centers have been leaders in using this technology. From student callers asking for an annual gift to United Way 211 operators guiding a person in trouble to the help they need, there are many ways speech-to-text can be utilized.

With annual fund calls, you no longer have to rely on the student's notes about the call to know how it went. You can capture the alum's reasons for giving or not giving. With 211 you can quickly determine if the right advice was given.

Speech recognition is sophisticated enough to distinguish callers, so you can apply your tone to both the caller and the receiver. Here is an example of a 211 call:

> Operator: 211, how may I help you?
>
> Caller: My family is hungry and we don't have any money.
>
> Operator: Where are you located?
>
> Caller: We are living in our car.
>
> Operator: Where are you now?
>
> Caller: Near the city park.

Operator: The Metro Food Bank is near you and open until 6pm.

Caller: Thank you.

Speech Recognition also includes the analysis of how words are spoken. This can reveal if a person is happy, angry, or sad. It can also infer if the people speaking are talking over each other. This could be used in conjunction with the transcription to glean insights into how callers are handling particular situations. You might have a caller who is very skilled with dealing with angry constituents while another one is not.

If you have any experience with Speech Recognition, you know it is not a perfect science. The first challenge is the audio quality. The second is the quality of the speaker or speakers. Finally, it is the ability of the speech recognition application to understand any words unique to your domain.

The audio quality can be addressed in two ways: the technology you use to record the calls and post-recording enhancements. Doing it right is far better than fixing what's wrong, so your investment in recording technology will pay off quickly.

It is hard, and in some situations impossible, to change the quality of the person speaking, especially the caller. With operators, you can work with them to have clearer pronunciation, but even here you are going to have good operators with thicker accents or imperfect grammar.

Some speech-to-text systems allow you to train the system to recognize a person's voice patterns. Dragon®, produced by Nuance, is one of those systems. They are licensing their technology to other companies, so we can expect this training capability to become more available.

Once you get past the quality of the recording and the people speaking, you face the final obstacle of what the application knows, or doesn't know, about your domain. Applications have hundreds of thousands of common words, but they do not have words unique to your domain. The application looks for a word it does know that is close to the word in your domain. For example, the Susan G. Komen organization could be changed to Susan G. Omen.

This is addressed with an ontology containing unique words in your domain. This forms the basis for a custom model you provide the application. Now when it has Komen in its vocabulary, it knows it is likely to be used with Susan G.

Experts say human accuracy is around 95%, so we miss one word out of every 20. IBM and Microsoft are in a race to reach this level of accuracy. As I write this, IBM has announced they have achieved 94.5% accuracy in a test using a standard testing set called SWITCHBOARD containing day-to-day conversations about such things as buying a car.

Using a more complex testing set (CallHome) with colloquial conversations between family members of a random set of topics, humans have a 93.2% accuracy rate, and IBM had 89.7%.

You could wait until parity, or even superiority, is achieved vs. humans or you could recognize having 100% of your calls analyzed at a slightly lower accuracy rate is far superior to having just a sample of calls analyzed. In the cognitive era, successful organizations will not wait for perfection to move forward. Accepting our current limitations helps make this much easier.

What are some of the benefits Speech Recognition can bring? A better experience for callers and operators is top on the list.

In fact, you will even be able to match up callers and operators based on what you know about each. This is already done with language, but now you could match callers and operators based on how comfortable the operator is with different personality types. In addition, operator training will be improved, and this will lead to better employee retention and constituent satisfaction.

As more of our technology becomes voice-driven, improving your organization's ability to understand and analyze those interactions will be become critical. Gone are the days when audio is driven just by phone calls. Constituents will be talking to your website and asking for information from your databases without touching a keyboard.

Your organization's mission will also be impacted as more and more information is brought in from the field, as people are no longer limited by access to their laptop and the speed of their typing. The doctor in the village being able to provide data in real-time as she speaks about a patient; the scientist describing river conditions for an environmental study; or the major gift officer providing rich details of the meeting they just had with a prospective donor are examples of how Speech Recognition is going to change the way we transfer information from our brains to the computer brains of our organizations.

From databases to the Internet, until now technology has been designed to inhibit voice communication. In the cognitive era, we get back to what we do best – talk.

CHAPTER

06

Inside the Science of the Cognitive Computing Brain

"The mind is like an iceberg, it floats with one-seventh of its bulk above water."

—Sigmund Freud

One of the hottest college degrees today is Cognitive Science. It is so new the definition of the degree varies widely from one institution to another. Johns Hopkins University describes it this way:

> Cognitive science is the study of the human mind and brain, focusing on how the mind represents and manipulates knowledge and how mental representations and processes are realized in the brain.

Conceiving of the mind as an abstract computing device instantiated in the brain, cognitive scientists endeavor to understand the mental computations underlying cognitive functioning and how these computations are implemented by neural tissue.

Cognitive science has emerged at the interface of several disciplines. Central among these are cognitive psychology, linguistics, and portions of computer science and artificial intelligence; other important components derive from work in the neurosciences, philosophy, and anthropology.

The "interface of several disciplines" does not do justice to what we are attempting to do. We are literally creating technology to mimic something we are still trying to fully understand – the human brain.

Cognitive science is where people with over developed left-brains are working with their counterparts on the right. Hopefully, they are both being managed by people who have balanced-brains.

To make Cognitive Computing successful, the breakthroughs will come from teams. You can see this in the platforms being offered by Google, IBM, and others. They all have places not just for the data scientists, but also for the subject-matter experts (SME). Without the data scientists it will all remain theoretical, and without the SMEs the data scientists will be lost in the vast frontier of how our mind works.

We are at the beginning of the cognitive era, and what we have today will feel like a flip-phone in the not too distant future. What we do have is the foundation on which this future will be built, and it starts with machine learning.

Machine Learning

Your current database is nothing more than a very smart filing cabinet. Instead of drawers, you have tables, and instead of folders, you have fields. It has no ability to think about the data, and only provides it when asked through a query by you.

Machine Learning is literally your computer learning iteratively from the data you provide it. 24/7/365 it is continuously looking for patterns and correlations. It is also able to change as new data is added. This means the insights provided will always be based on the latest information available. While this is a truly extraordinary capability, it is also why you need to be careful how it is implemented.

I learned this in 2004 as we brought wealth screening data to the software-as-a-service world. We added a real-time connection from our stock data to stock prices enabling wealth to be recalculated as prices went up and down. The problem was this meant clients who created "top lists" based on the value of wealth saw their lists change constantly. This was disconcerting, and clients were not happy. We solved the problem by having two types of lists: dynamic and static. With a static list, the wealth data still changed, but the prospect names on the list remained the same.

Today you could use the changing wealth data to create dynamic invitation lists where people are invited to an event up to a certain date if their wealth qualifies. A more serious situation is driverless cars. The computer has to constantly adapt to changing conditions, and the decisions literally can be life or death.

The lesson is to only make a direct connection of your insights to your operations when you are confident the decisions (including the bad ones) are superior to what can be done without the Machine Learning.

One of the challenges, and biggest opportunities, of Cognitive Computing is harnessing the power of continual learning. You will need to take this into account as you integrate cognitive tech decisions into your processes.

There are four types of Machine Learning, each with their own algorithmic approaches:

> **Supervised Learning**: Just as it sounds, a person supervises the process. You provide a training set of correct answers, and one of incorrect answers. An example is identifying unhappy donors based on incoming emails. You would first identify emails where you know the donor is unhappy, and then identify emails where they are either neutral or happy. Supervised learning is best suited for use cases where classification is straightforward, and you are able to create positive and negative training sets.
>
>> **Algorithms**: Decision Trees; Back Propagation; Nearest Neighbor; naive Bayes; Linear Regression; Logistic Regression; Support Vector Machines; and Neural Networks.
>
> **Unsupervised Learning**: Here the machine draws inferences from data without a particular answer in mind. There are no teachers and no correct answers. In our email example, you would just provide the machine with a large group of uncategorized incoming emails.

The machine breakdowns the emails into what it perceives as logical groups. These groups will be unlabeled, so the final step will be for you to label them. Unsupervised learning could be the best approach for this use-case because it is unlikely you have categorized your emails into happy and unhappy donors, so doing that would require a lot of time. Another advantage is you will be presented with other email clusters that might be valuable for other use cases.

Algorithms: K-means Clustering; Association Rules; and Apriori.

Semi-supervised Learning: This technique is used when labels are only available for a portion of the data. You can think of it as splitting the difference between Supervised and Unsupervised Learning.

Algorithms: Self-Training; Generative Models; S3VMs; Graph-Based and Multiview.

Reinforcement Learning: Algorithms built with Reinforcement Learning are not given specific goals, but rather learn through trial and error. They are rewarded (reinforced) for doing the right thing and punished for doing the wrong thing. You can think of how a dog is trained or how you learned to get through the levels of a game faster and faster. This style of learning is not suited to every problem, and it requires a great deal of computational power. As the availability of computational power goes up, and the cost goes down, this type of learning is gaining more traction.

Algorithms: Q-Learning; Temporal Difference; and Deep Adversarial Networks.

A simpler way of thinking of the four processes is Labeled; Unlabeled; Partially Labeled; Labels Not Applicable. In Supervised Learning, you are labeling things correct or incorrect while with Unsupervised Learning you have provided the same data, but without a label. It labels the data on its own.

An advantage of Unsupervised Learning is once you have your algorithm up and running, you have created an active, rather than passive, learner. The algorithm will adapt to new input better than a Supervised Learning algorithm. Reinforcement Learning takes this to another level.

Let's say your organization focused on decreasing smoking among teenagers, and when you did your Supervised Learning you did not include e-cigarettes. Unsupervised Learning would have created a classification system to identity what people look like when they are smoking, and would just create a new cluster of teenagers with this type of cigarette.

Here is how Facebook delivers your news feed using machine learning:

Facebook's News Feed uses machine learning to personalize each member's feed. If a member frequently stops scrolling in order to read or "like" a particular friend's posts, the News Feed will start to show more of that friend's activity earlier in the feed. Behind the scenes, the software is simply using statistical analysis and predictive analytics to identify patterns in the user's data and use those patterns to populate the News Feed. Should the member no longer stop to read, like or comment on the friend's posts, that new data will be included in the data set and the News Feed will adjust accordingly.

http://whatis.techtarget.com/definition/machine-learning

One more term to know is the Bayes Rule. This comes from the Bayesian theorem, named after 18th-century mathematician Thomas Bayes, who introduced the concept of probability to prediction. The Bayes Rule is the confidence in a probability of an event based on prior knowledge of conditions that might be related to the event.

Let's look at how we could apply the Bayes Rule to wealth screening:

- Let's assume your organization defines a major gift prospect as having a gift potential of $100,000 or more.
- 5% of your donors have the capacity to be a major gift prospect (therefore 95% of donors do not have the capacity to be a major gift prospect).
- 80% of the donors who have made a major gift were correctly rated by wealth screening (therefore 20% of the people who gave a major gift were identified as having a lower capacity by wealth screening).
- 10% of donors identified as a major gift prospect were in fact not capable of making a $100,000+ gift (therefore 90% were correctly identified as not having major gift capability).

This data can be used to predict the probability a major gift rating is correct.

	Major Gift Prospect (5%)	Not a Major Gift Prospect (95%)
Major Gift Prospect	Correctly Identified 5% x 80% = 0.04	Incorrectly identified 95% x 10% = 0.095
Not a Major Gift Prospect	Incorrectly Identified 5% x 20% = 0.01	Incorrectly identified 95% x 90% = 0.855

To get to our probability we add the correctly identified 0.04 and the incorrectly identified 0.095 to get the total possible prospects identified as having major gift capacity, which is 0.135. We now take our correctly identified number of 0.04 and divide it by 0.135 to get our probability of 29.6% that any person rated as a major gift prospect actually has the capacity to give $100,000+.

You might be disappointed the number is not closer to 100% (or at least the 80% you thought it was), but keep in mind you now have a 1 in 3 chance rather than a 1 in 20 chance (5% give at the major gift level) by randomly selecting prospects. This is also why you need to have a prospect research professional to close the gap between probability and reality.

The Bayesian approach to data science is especially useful when you are working with a dataset involving a high degree of error. Fundraising and mission delivery are both rife with error given how much they involve human interactions.

If you are a data scientist, then you most likely already knew about the importance of factoring in probability. If you are not, ask your data scientist how they are applying the Bayes Rule to their analysis, and they might (happily) do a double-take.

Despite all of the evidence many of the algorithms produced by Machine Learning perform better than humans, there is still strong resistance to using them. Dr. B. J. Dietvorst has completed a number of studies on the subject. His latest working paper, *People Reject Superior Algorithms Because They Compare Them to Counter-Normative Reference Points*, looks at the phenomenon of how people compare algorithms to their own performance goals even when those goals are unrealistic.

Machine Learning is especially vulnerable to this type of comparison because of its iterative nature. If we see this as a competition, then we will claim the algorithm has failed before it has learned, and even when it does, we will claim it is still not as good as we could be.

Not unlike the person on a horse laughing at an early car puttering along a dirt road, we try to see ourselves as superior to whatever is new. Just how far behind are we willing to let our organizations fall before we realize embracing this technology is what will carry us forward?

Deep Learning

I said this was not going to be a "how-to" book, so I am going to keep this as understandable as possible for folks without a computer science background. This means those with one will recognize I am leaving out a lot of detail and some of the more advanced ways deep learning can be implemented.

Deep learning is doing nothing less than trying to mimic how our brain functions. That means we are trying to simulate the communication of 100 billion neurons capable of doing everything from reading this book to discovering a cure for a type of cancer.

In a human, a neuron is a cell. In a computer, it is a processing unit. These processing units are divided into input and output neurons with the input being the data used to determine the output, which is the answer. The input neurons are layered from the bottom up with the output neurons being on the top. This layering is called a neural network

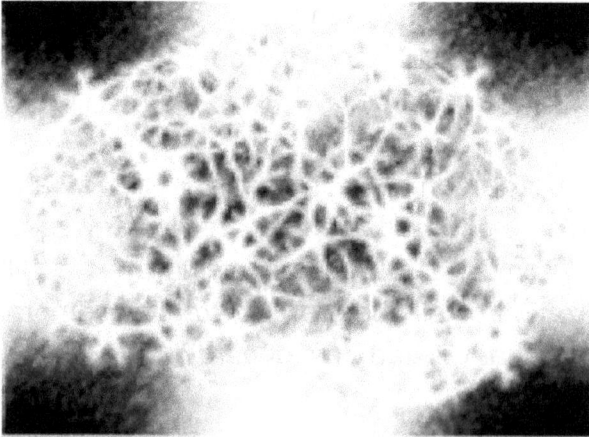

We will refer to the computer neurons as artificial neurons (AN) to distinguish them from their human ancestors. Each AN has one or more inputs and is given a specific weight. The

ANs are able to communicate with each other just as our neurons do.

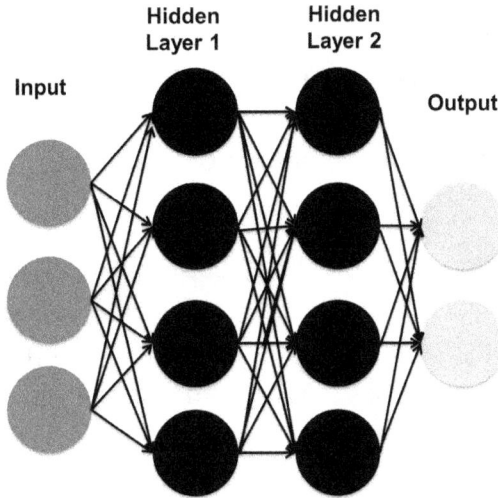

Analyzing what is in a picture is a good way to understand how this works. Let's say you want to be able to recognize your school's mascot. You provide the system hundreds if not thousands of pictures of your mascot in various venues from sporting events to selfies with fans. This is your positive training set. Now you provide mascots from other schools, and these are your negative training set.

Each photograph will be broken down into sections and provided to an AN. The system will learn different aspects both of the mascot and also about the settings where the mascot is likely to be (a basketball arena for instance). You can start to imagine how the layering or neural network comes into play. The computer is building to the answer as each section reports its results.

This is being used successfully to analyze everything from dog breeds to mammograms. You can see how an environmental group could use it to monitor the health of a wildlife preserve or a health organization to see who is wearing their event t-shirts on social media.

Google has used Deep Learning to create headlines from article text and to even reply to emails. Your marketing team could use these same techniques to create the best subject line for your next email campaign or to reply to standard email questions.

Another exciting area for Deep Learning is time-series data. This is where your data is analyzed in the context of when it appeared. Netflix and Amazon use it to recommend movies and items to purchase. You can use it to increase the accuracy of your campaign forecasting or to match the interests of your donors with different aspects of your mission.

Resource allocation is another huge opportunity for cost savings. Look at both your mission and business operations to see where you might apply Deep Learning to steward your precious resources more effectively.

It will not be long before organizations will have an organizational brain thanks to Deep Learning. The organizations embracing this capability will leapfrog their peers as they move from being a data-driven to an insight-driven organization.

For those of you who just can't wait to learn more, I suggest looking at TensorFlow™, developed by the Google Brain Team. It is open-source, and there are plenty of great examples of how to put it to use including photo recognition—https://www.tensorflow.org/tutorials/image_recognition.

Internet of Things

©2017 Max Lawson

Two air disasters in the last few years demonstrate what Internet of Things (IoT) technology can do for us, and what not having it has cost us.

The first is the disappearance Malaysia Airlines Flight 370. Other than last communications, and a few radar and satellite pings, we have no idea what happened to the plane or where it is today. A recent report issued by Australian investigators has stated it is almost inconceivable the aircraft has not been found.

The second is EgyptAir flight 804, which crashed in the Mediterranean. This plane transmitted data in real-time revealing a rapid loss of control, with alarms and computer-system failures in the seconds before the plane was lost from radar. They didn't have to wait until they found the wreckage and the black boxes, which they did quickly.

The Internet of Things connects devices directly to data repositories. While you may not have a 757 to track, you may well have access cards; smart energy devices; RFID chips on event bibs; and branded Wi-Fi.

Colleges can use access cards to see how students, and faculty, are using facilities. This can help with safety, personnel deployment, and energy use as you see where and when people are entering and leaving buildings on campus. Facilities can use data from smart meters to help lower costs by managing temperature based on actual usage rather than guesses.

Taking a page from Disney®, you could have bracelets at your next event. The bracelets can allow people to attend particular functions. You will be able to see how people are experiencing your event. Unlike Disney that wants to know your name, you can anonymize the data. The big insight is to see which aspect of your events are effective, and which are not. Seeing not just how many people came, but how many stayed and for how long is invaluable for planning your next event.

If your organization has people in the field, this technology can help keep track of people and equipment. If trucks are involved, such as with Goodwill®, then IoT can be used just as it is with UPS® to help plan routes and reduce costs.

When you offer free Wi-Fi, consider requiring users to give you their email. This will enable you to see who is using your facilities, and it is also a great opportunity to put your message in front of people.

The use of IoT to further the mission of organizations is the most exciting use of this technology. An example of this is the Jefferson Project at Lake George, a partnership between Rensselaer Polytechnic Institute (RPI), IBM Research, and The Fund for Lake George to monitor the health of Lake George using IoT and analytics. Using over 40 sensors, the project continually monitors the health of Lake George, providing streaming data far superior to traditional sampling techniques.

The Jefferson Project at Lake George also demonstrates the importance of integrating analytics and Machine Learning with IoT. We have all heard the phrase "fire hose of data." With IoT, it is more like a Niagara Falls of data. Not only can you setup as many sensors as you want, those sensors will be delivering data every second of every day.

With all of this incoming data, a solid platform for analytics and Machine Learning is essential in order to ingest; transform; and analyze in real-time. Fortunately, this type of platform is readily available from a variety of vendors. The cost is decreasing almost daily, both because of improvements in technology and competition. Another important factor with cost is adoption. Just as with any other product, the more that is sold the less it will cost any one customer.

Based on the success of the RPI/IBM partnership, it is not hard to imagine every environmental project going forward will make use of IoT. Being able to monitor the quality of an environment without the constraint of human involvement will dramatically increase our knowledge and decrease costs. We can then focus our expertise on the interpretation of the data and the actions which need to be taken based on the insights.

The Nature Conservancy (TNC) is tracking bird migration to understand all of the places needing protection:

> "We tend to think about biodiversity as a static thing," says Jorge Brenner, a TNC marine scientist based in Texas. In fact, he says, animals move through space and time, urged on by the need to breed or to feed. And, as many scientists know all too well, they can be

surprisingly hard to follow as they flit around a forest or dive beneath the waves. But that is changing.

Today, technology makes it easier than ever to follow animals large and small. Satellite and GPS monitoring tags get smaller and lighter every year, thanks to battery improvements and solar-powered options, allowing scientists to trail ever more petite species while minimizing animal disturbance. Newer tracking systems are able to forgo satellite monitoring and instead rely on cellular networks, which can handle more data and provide a critter's coordinates at more frequent intervals.

https://www.nature.org/magazine/archives/animal-tracks.xml

IoT offers the promise of identifying problems from the mundane to those of life and death. It may be your car's taillight is going to go out in a week, so replace it now. It may be an issue on the flight you were about to take, which resulted in a call for maintenance and, ultimately, averted disaster.

We do not lack for social and environmental fire detectors. What are in short supply are smoke detectors, telling us what we need to know before it is a crisis.

AI – Artificial or Augmented Intelligence?

AI is what captured our imagination, not Machine Learning or even Deep Learning. Neither sounds like much fun. On the other hand, a machine with the intelligence of a human makes you go to the movies to see what happens.

The majority of the movies have portrayed a dystopian future where AI either attempts to kill us (*War Games*); uses what it knows about us against us (*Minority Report*); or takes over the world (*Terminator*). Not surprisingly this had led to a bit of suspicion as AI has moved from the theatre to our living rooms.

My introduction to AI was the movie *2001 A Space Odyssey*, Stanley Kubrick's masterpiece based on Arthur C. Clarke's novel of the same name. In the film, a mission to Jupiter includes a computer, HAL 9000, which has the following capabilities:

- Speech
- Speech recognition
- Facial recognition
- Natural language processing
- Lip reading
- Art appreciation
- Automated reasoning
- Behavioral interpretation including emotions
- And it plays chess

Sound familiar?

Spoiler alert for any of you who have not seen the movie or read the book - HAL malfunctions and things go horribly wrong. Many famous lines came out of the movie and it was not lost on me as a young child, the astronaut was named

Dave (although there are so many of us, the odds were in my favor). Years later I would use sound clips from the movie to replace various MS-Windows sounds. For errors I used, "I'm sorry Dave, I'm afraid I can't do that."

Clarke and Kubrick did not invent AI. The term *Artificial Intelligence* was first used at a conference at Dartmouth College in 1956. Many of the concepts we are now seeing come to fruition have been around for decades. The problem was not so much the science; it was the low level of computer power available and the limitations and expense of storing data.

Starting with IBM's Deep Blue beating chess champion Garry Kasparov in 1997, the field of AI started to gain a new momentum. The power of computers grew exponentially even as the cost of computing power dropped. Hadoop revolutionized how data was stored, and by 2011 IBM Watson was winning *Jeopardy!*.

What exactly is AI? John McCarthy, a Professor at Stanford University described it this way in 2007: "It is the science and engineering of making intelligent machines, especially intelligent computer programs. It is related to the similar task of using computers to understand human intelligence, but AI does not have to confine itself to methods that are biologically observable."

The last part of his description is especially intriguing. We believe we are reimagining our brain as a computer, when in fact we are now partnering with computers to create something currently not observable.

Despite all the hype, AI is in its infancy. It is doing amazing things, but we are not unlike the Wright Brothers flying over the sand dunes trying to imagine how their invention would change the world. For all their brilliance, I doubt they foresaw traveling to Europe in a matter of hours while deciding whether you wanted the red or white wine.

Another challenge with AI is the term is now used too often by companies as a marketing tool rather than an actual technological advance. This leads to people being disappointed as they try out what they think is AI, and find it lacking. I call this AI-Lite.

A good way to look at AI is stages. The first being where it is able to take over a task, such as customer service interactions using a chatbot; the second is understanding a domain, and being able to not only make predictions but also prescribe solutions as well as humans; and third is becoming as smart as or smarter than humans.

We are well underway with stage 1, and we are seeing more and more happening in stage 2. We are on the path to stage 3, but it is far down the road (at least as of now).

Alan Turing, who created the fundamental architecture for modern computing with his Turing Machine, created the Turing Test to answer the question, "Can machines think?" The test involved a human and one or two partners who the human knew could be a computer. Using only text, a series of questions were posed by a human, and the computer and human(s) would answer. If the human could not distinguish the computer from a human, the machine was said to have passed the test.

You can see why *Jeopardy!* was not a random game chosen by IBM. It is perhaps the ultimate Q&A (or A&Q) game.

In 2014, the first computer program to officially pass the Turing Test (If a computer is mistaken for a human more than 30% of the time during a series of five-minute keyboard conversations it passes the test.) was designed by Vladimir Veselov. His program, called *Eugene*, simulates a 13-year boy. Thirty-three percent of a panel of 30 judges were convinced it was a human.

The chatbot industry was born at that moment.

So far we are measuring AI by how smart it becomes relative to us. Intelligence is certainly part of who we are, but it is far from all we are. The real question is can AI become Sentient. This is stage 4:

> **Sentience** is the capacity to feel, perceive, or experience subjectively. Eighteenth-century philosophers used the concept to distinguish the ability to think (reason) from the ability to feel (sentience). In modern Western philosophy, sentience is the ability to experience sensations (known in philosophy of mind as "qualia"). In Eastern philosophy, sentience is a metaphysical quality of all things that requires respect and care.

https://en.wikipedia.org/wiki/Sentience

While to many this is the ultimate sci-fi fright story, I actually am more concerned about the period between AI being smarter than us, yet not possessing the ability to comprehend (or more importantly care about) the consequences of its actions. For humans, this is called a sociopath.

How would you teach a computer to understand consequences? In 1998 Richard S. Sutton and Andrew G. Barto co-authored *Reinforcement Learning: An Introduction*. They proposed for computers to be able to learn on their own they need to interact with the environment as we do. This was based on decades of research into adaptive learning, and especially the work of the late A. Harry Klopf at the University of Massachusetts.

> "The learner is not told which actions to take, as in most forms of machine learning, but instead must discover which actions yield the most reward by trying them."

This type of machine learning enables a computer to go beyond answering a question or performing a repetitive task. It can now move on to problems with longer timelines requiring adaptation to external forces such as the scheduling of transportation, a robot moving in a hospital, or playing the game Go.

An important aspect of Reinforcement Learning is Temporal Difference Learning (TDL). If you are intrigued and want to know more, read Sutton's work on the subject. For the purpose of this book think of TDL predictions as being a co-opetition between different algorithms, where the competitors are all trying to win yet are also in constant communication to help each other.

Each algorithm is like a child in a pre-school, unable to figure out how to put the shaped blocks in a container with corresponding shapes on the outside of the container. She watches another child do it correctly and then mimics the process. Then, as one of my sons figured out, the first algorithm realizes you can open the top of the box and just place all the blocks inside. Not what the manufacturer had in mind, but if the problem is how to get the blocks in the box the fastest, he wins.

While you can start to envision how AI will one day become sentient or pretty darn close, what we are actually starting to see is a move away from the term *Artificial Intelligence* to *Augmented Intelligence*. It is both recognition of how the fear of what AI might become has slowed adoption, and an acceptance of the reality AI is still far from being capable of doing many of things people fear it might do.

Augmented is meant to convey the technology's purpose is to help rather than replace humans. It is the difference between auto-pilot in a plane and a pilotless aircraft. Pilots don't leave the cockpit (or at least not all of them) when they engage auto-pilot even though on long-haul flights they are not actually flying the plane most of the time.

A number of years ago while I was earning my first million Delta® miles, the pilot informed us they needed to test the auto-pilots ability to land the plane. We were on a Lockheed 1011, the plane with a tale engine looking like it was stuck on, and it always felt loose as the overhead compartments moved with any bumps. The auto-pilot kicked in as we approached Orlando, and very quickly started compensating for the wind. Then it began over-compensating taking us right and left and up and down. Not by a lot, but enough to put the "this doesn't seem right" face on all of us.

After about five minutes, the pilot disengaged the auto-pilot and came on the speaker to say, "Well, you probably guessed it didn't pass." Thank goodness the plane was augmenting and not replacing the human in the equation.

With a customer service chatbot, a company may want it to do all the work, yet there will be times when a customer still needs to speak with a representative. The best chatbots will have the intelligence to know when a human is required, and also which human to engage with the customer.

The medical profession is doing a lot of work with Augmented Intelligence. A doctor can use AI to read every medical journal and provide a summary of the relevant information for her to read. The doctor can follow up with more research on a particular subject, and will most likely still read the journals most pertinent to her specialty. The doctor welcomes the information and keeps her prerogative to make decisions based on it.

A researcher in the field can use AI to remotely query whatever data they need using voice commands. Data can be sent back to the AI in real-time, and feedback is received instantly. This intelligent back-and-forth enhances, rather than replaces, the researcher's capabilities.

IBM has recognized the debate about the implications of Cognitive Computing is slowing its adoption. Recently they introduced three principles to guide their use of the technology:

> **Purpose:** The purpose of AI and cognitive systems developed and applied by the IBM company is to augment human intelligence. Our technology, products, services and policies will be designed to enhance and extend human capability, expertise and potential. Our position is based not only on principle but also on science. Cognitive systems will not realistically attain consciousness or independent agency. Rather, they will increasingly be embedded in the processes, systems, products and services by which business and society function – all of which will and should remain within human control.
>
> **Transparency:** For cognitive systems to fulfill their world-changing potential, it is vital that people have confidence in their recommendations, judgments and uses. Therefore, the IBM company will make clear:

- When and for what purposes AI is being applied in the cognitive solutions we develop and deploy.

- The major sources of data and expertise that inform the insights of cognitive solutions, as well as the methods used to train those systems and solutions.

- The principle that clients own their own business models and intellectual property and that they can use AI and cognitive systems to enhance the advantages they have built, often through years of experience. We will work with our clients to protect their data and insights, and will encourage our clients, partners and industry colleagues to adopt similar practices.

Skills: The economic and societal benefits of this new era will not be realized if the human side of the equation is not supported. This is uniquely important with cognitive technology, which augments human intelligence and expertise and works collaboratively with humans. Therefore, the IBM company will work to help students, workers and citizens acquire the skills and knowledge to engage safely, securely and effectively in a relationship with cognitive systems, and to perform the new kinds of work and jobs that will emerge in a cognitive economy.

While these are ostensibly for IBM, they really are guiding principles for every company and organization using cognitive technology. Your organization needs to have their principles front and center as you embark on your cognitive transformation.

A final thought on Artificial vs. Augmented Intelligence: Given this is the Cognitive Computing era, I support *Cognitive Augmentation* as the overarching term. This allows for Augmented Intelligence; Augmented Machine Learning, and so on. It also positions the technology where it belongs, as a cognitive partner to help humans survive and thrive.

07

Cognitive Analytics – Getting Beyond the Dashboard

"We may say most aptly that the Analytical Engine weaves algebraic patterns just as the Jacquard loom weaves flowers and leaves."

—Ada Lovelace

Traditional analytics provided the illusion we were doing in-depth analysis of our data. Dashboards gave leadership a way to consume this incomplete, and static, analysis. The two combined to keep everyone happy until someone figured out the insights were misleading, and often-just plain wrong.

We are all familiar with Key Performance Indicators (KPIs), those measurements of activity and outcomes such as the number of people your organization has served or how much money you have raised. KPIs are fine as long as you don't care why the numbers are what they are.

Scoring of constituencies has been around since the early 80s. Today we have scores for just about everything from wealth to interests to how likely you are to graduate from a university. Organizations have successfully used these scores to better segment their files, and the results have been far superior to the old follow-your-gut or follow-your-peer approach.

The problem comes when you dig deeper into the scores and realize while they are better than doing nothing, they are unable to capture the true nature of your constituency. This is a result of the methods used in traditional analytics such as omitting people at the extremes (outliers) and having strict rules on what data can be used in the model.

Often organizations are seeking to look for people at the extremes: from very wealthy people to mission beneficiaries at the margins of society. The data leading you to these people may not be available on a lot of the records in your database. How many have a billion dollars in stock or just lost their job; have four young children; and has a close relative with a substance abuse problem.

Cognitive Analytics acts not like a statistician, but like you do when given information about an individual. You don't ignore a fact because it doesn't fit your model. In fact, those are often exactly what draws you to the person.

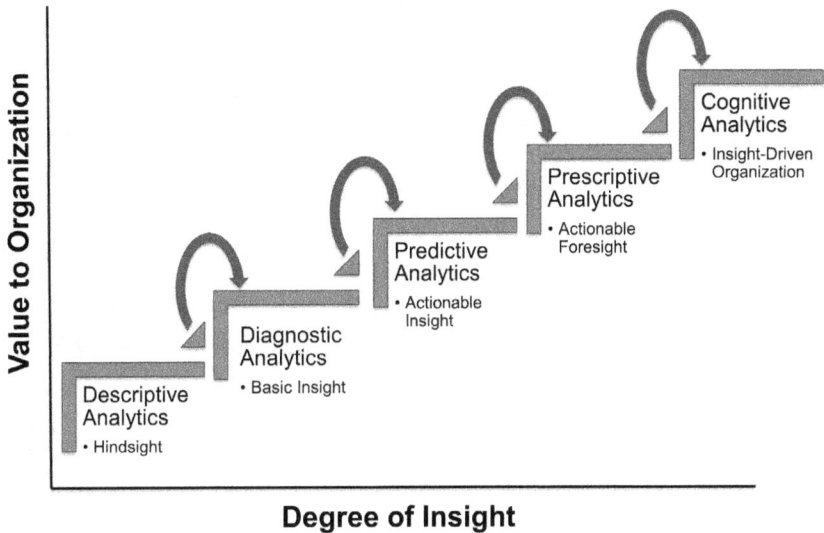

Degree of Insight

This doesn't mean all of the traditional analytics are thrown away. Think of them as part of the foundation of Cognitive Analytics. When I am asked by someone who has done data science using classic tools whether they can learn the skills for Cognitive Analytics my answer is an emphatic yes! I tell them the challenge will not be learning, but rather unlearning what they have known about the limitations of analytics. It is okay to have a problem requiring a lot of computing processing power. Not a problem if you want to analyze terabytes of data. The limit will be your imagination going forward. Let's take a look at where you can go with Cognitive Analytics.

Cluster Analytics

Predictive analytics has been hobbled by its reliance on structured data and classic regression analysis. An example is RFM (Recency; Frequency; and Monetary Value). Simply, what this will tell you is who has recently made a gift; gives

119

frequently; and gives you a lot of money. In regression analysis, you could use these three variables to create a score, so that your *best* givers are on top. You could then take that score and add other variables such as age; gender; and location in a query to produce a list of prospects.

In Cluster Analytics, you are able to take each element of the RFM and create clusters based on both structured and unstructured data elements. This would provide insight into which types of donors have given to your organization in the last year; given a number of times; and given your organization a high amount of money. Unlike a query where you decide the variables that create the group, with Cluster Analytics, those variables are based on the reality in your constituency. You may find clusters where age or location is irrelevant.

One example of Cluster Analytics I like is a school dance. Here are *some* of the variables you could use:

- Girls
- Boys
- Transgender people
- Dancers
- Non dancers
- Girls who want to dance
- Boys who want to dance
- Transgender people who want to dance
- Girls who want to dance with boys
- Girls who want to dance with girls
- Girls who want to dance with transgender people
- Boys who want to dance with girls
- Boys who want to dance with boys

- Boys who want to dance with transgender people
- Transgender people who want to dance with girls
- Transgender people who want to dance with boys
- Transgender people who want to dance with transgender people
- Girls who don't want to dance
- Boys who don't want to dance
- Transgender people who don't want to dance
- 12 & under
- 13 year olds
- 14 & older
- GPA 2 and under
- GPA 2.01-3
- GPA 3.01-4
- People who like the music
- People who don't like the music
- People who are sick
- People who are healthy

In Cluster Analytics, you can create groupings of these variables far beyond what classic predictive analytics would create: girls and boys who have the same GPA range, and don't like the music, but want to dance; and 13-year-old girls who want to dance, and like the music, but didn't dance, and so on. Some of this information will not be available unless you did a survey, but I included it to get you thinking about what you might ask your next event attendees in a follow-up survey.

Unless you are in middle-school, this might just be bringing back uncomfortable memories, so let's look at ways you can use it.

> Grateful Patients: Incorporate patient satisfaction survey data, demographics, and other HIPAA compliant information, with their giving history.

> Alumni/ae: Incorporate social network interactions; verbatim survey responses; and email campaign data with giving history.

> Member Organization: Incorporate event attendance, RSVP with regrets, membership activity with giving history.

Clusters are also critical to Journey Analytics. The clusters provide groups of constituents to follow through your organization.

Traditional analytics is just not capable of dealing with the multi-dimensional data found in the real world. Cluster analytics doesn't see any data as "outliers," rather it seeks to create clusters which incorporate the outliers.

My first experience with the limitations of traditional analytics was seeing how scoring models were being built to segment donor files by wealth. I approached some of the experts building these scores with the more precise data my company was providing on stock holdings, and other assets. I was told this data could not be incorporated into their models because it was only available on a relatively few people.

I pointed out their scores often were well below what the facts from our data clearly indicated. Again, they said it would hurt their overall model to have this higher quality, but lower quantity, data in the mix.

Now with cluster analytics, these so-called outliers can be grouped into a cluster rather than either taken out of the analysis or having the wrong score. Think about adding personality; tone; and sentiment to the mix and you can see how without cluster analytics we would not have a way to turn them into the insights we need to understand how our organization is really doing.

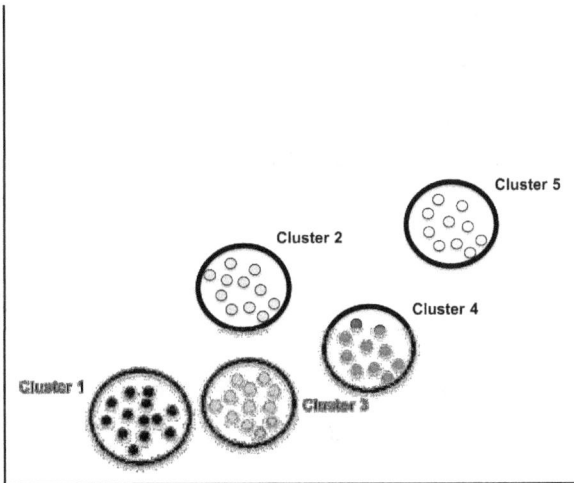

Visual Recognition – Understanding What We See

Pictures are not only worth a thousand words; they also exponentially increase your storage needs. This is one of the reasons images have not been stored in the past. The other reason is while they were nice to have, there was no way for an organization to analyze them.

Now, thanks to cheap storage and Cognitive Computing, we can incorporate Visual Recognition into our repertoire. How do you get started? You first need to decide what visual data is important to you. It might be as simple as your logo or as complex as buildings on your campus. Once you have decided what is important, create the training sets.

Let's take your logo: you provide a couple of hundred different ways your logo has been displayed from clothing to coffee mugs to signs at an event. You would then provide a similar number of logos of other organizations. You would do the same with buildings on your campus.

You are telling the cognitive platform what is and what is not what you want to find. Once you are comfortable with the accuracy, you can begin analyzing sources of visual data, such as social media. Imagine being able to identify people who wear your logo or post pictures of your campus. Associating yourself with a brand says a lot about your feelings towards that brand.

Over 350 million photos are uploaded Facebook® every day. As Instagram® becomes a more important platform, photos will take an even more important role in social media. It's a lot easier to show you had a great time at an event by uploading a photo of you and friends with smiles, rather than typing, "We had a great time...."

A computer *sees* images very differently than we do. Rather than looking at the image as whole, it breaks it down into individual sectors. It learns what is in each sector, and what different combinations of sectors create. In this example I would have a training set with different pictures of GoodBot™, and in time it would come to learn what her characteristics are such as the mechanical wings, ears, and the shape of her eyes. Using this information it would be able to distinguish her from similar images like BadBot™ and also find her in pictures with backgrounds it had never seen before.

Metadata is another source of invaluable insights from photos. A photo's metadata allows you to learn when the picture was taken, and even the location. Using this information, you can literally map where in the world your organization's logo is being posted over time.

A very practical use of this is to monitor activity during a fundraising campaign or event. Is your social media campaign working? Is your event going viral? That can all be seen with Visual Recognition.

You can also monitor how your brand is being displayed year over year. This can be incorporated into an overall Brand Score. You can then layer in clusters to see which groups of constituents are more or less engaged. You might be hitting it out of the park with Boomers, but going bust with Millennials, or maybe it is certain parts of the geographic area you serve are represented more so than other locations.

What about people? Your leadership, professors, and key staff might be worth monitoring. In this case, you do the same thing you did with your logo and buildings, except this time it is different pictures of the people you want to find.

If your organization is in a field where there are images closely related to it such as the environment or the arts, you can teach the platform to recognize those images. Knowing your constituents post images of nature or trips to museums provides insight into their interests.

While it is great to see your logo, what about looking for the logos of organizations similar to yours? It may pain you a bit to see your donor wearing someone else's swag, but in the real world people rarely support just one organization. Suck it up like Starbucks® does and accept not every coffee drinker is going pay $5, and not every person who pays $5 is going to only do it at Starbucks.

Have you ever wondered if people enjoyed an event your organization hosted? You could use Microsoft's Emotion API to analyze pictures from the event, and be able to discern anger, contempt, disgust, fear, happiness, sadness, and surprise on the faces of your attendees. Imagine then adding the context of time to see whether different aspects of your event were received the way you wanted them to be.

The potential uses for Visual Recognition for mission work are just beginning to be explored. Intel® is using it to help the National Center for Missing and Exploited Children (NCMEC) CyberTipline, which is a reporting mechanism for people to report incidents of child sexual exploitation. Their task is daunting:

> Search tens of millions of images in an effort to identify children and suspects; to associate images with network addresses or domain names; and to distinguish new images from known ones. There are

about 460,000 images of missing children in the NCMEC database, and last year NCMEC received more than 8 million reports of possible child sexual exploitation.

NCMEC only has 25 analysts to handle this tremendous amount of information. Now they have a cognitive partner to mine of all the leads and put in front of them the most promising. The value of finding more children and putting more predators behind bars is incalculable.

Food insecurity is growing around the world, and with the exponential growth in population promises to be one of our primary challenges moving forward. Being able to increase crop yield, while reducing pesticides, is a goal for all food producers. Visual recognition has been combined with drones to analyze photos of crops to look for disease and pests.

NatureSweet®, using technology from an Israel-based company Prospera®, installed 10 cameras in a greenhouse to continually monitor tomato plants. Using visual recognition, it is able to identify insect infestations and dying plants. NatureSweet believes yield could eventually be improved by 20%. They are also working on identifying the optimum time to harvest the tomatoes.

Environmental groups will train their visual recognition systems to identify problems in the ecosystems they monitor. They can enlist volunteers to upload pictures of at-risk areas.

Our ability to capture images is now almost limitless. Combine that with Visual Recognition, and the power of image insights to transform your organization's operations and mission is limited only by your imagination.

Journey Analytics

Creating donor experiences was at the center of our work at The WOW! Institute, an experiential learning event I co-founded with Jay Goulart. Jay is a master at designing journeys for donors leading to life-long relationships. For 4 days each summer from 2001 – 2007, we brought fundraisers, leadership, and philanthropists together to discover how they could become journey designers.

A recurring question was, "How do we scale this?"

We incorporated what then were the latest digital engagement tools, but we understood a great deal of the activity was going to be focused on the higher-level donors where relationship fundraisers could personally deliver the experiences. Now the combination of omni-channel marketing tools, and Cognitive Computing, make it possible to deliver meaningful experiences at scale.

Journey Analytics, one of the most exciting cognitive tools, is what brings Big Data into the marketing mix. With Journey Analytics, you are able to fully understand how people move through your organization, both physically and virtually, which provides deep insights into what is working and what is not.

The first step in designing your journey map is to identify every touch point. From emails to events to social networks, you want a complete list of all the ways a constituent might interact with your organization. With this in hand, make sure your Insight Reservoir has data from all the touch points. Next, create golden constituent records across all the touch points.

At this point, you can take an individual through the journey over any given timeline. While this is valuable to a major gift officer or a caseworker, the organization will want to look at how journeys cluster. Here, another element needs to be added – outcomes. With fundraising, it might be total giving, time on the file, or median gift. For your mission, it might be finding employment, a home, or graduating.

Another dimension of Journey Analytics is the ability to see how different journeys are impacting different metrics. For a small organization you might organize your metrics this way:

- Total Giving by Levels (Under $50; $50 - $99; $100 - $499; $500 - $999; $1,000 - $4,999; $5,000 - $9,999; and $10,000 and up)
- Largest Gift (Under $100; $100 - $999; $1,000 - $4,999; $5,000 - $9,999; and $10,000 and up)
- Number of Years of Continuous Giving (1; 2; 3; 4; 5; 6+)
- Number of Gifts (1-3; 4-6; 7-10; 10+)

What you typically find is there is no one journey yielding the highest levels within each metric. One journey performs best with a certain cluster for Number of Years of Continuous Giving, while that same cluster under performs for Largest Gift.

Armed with this insight, you ask the most important question – why?

Perhaps you have a good program for keeping mid-level donors engaged, but your major gift program is not resourced properly to take people closer to their giving potential. Or maybe there is something deeper going on.

This is where qualitative data comes in to provide the color commentary for the journey. Hidden in your gift officer notes, survey responses, and social posts and comments are insights into the drivers of your results. Perhaps you just have not made the case for higher levels of support, a problem I have seen at many organizations with mature low- and mid-level programs, but an under-resourced major gift program.

On the mission side, think about a university being able to factor in where the student lived as they look at their academic journey. I was speaking with a student entrepreneur in my work at Domi Station, a business incubator in Tallahassee I co-founded, and asked about life on campus. She told me about the mold in her dorm, and how it made her sick. She had to throw away all her clothes at the end of the year as did her classmates. Imagine being able to take this type of data into account when analyzing students' absence rate, retention, and even GPA.

For school fundraisers, maybe mixing in where your alums lived while attending your school might give you an import data point as you seek to understand why they are, or are not, giving.

Journey Analytics is the perfect complement for the Theory of Change (TOC). For those not familiar with TOC, it is a methodology for analyzing the effectiveness of an organization's programs. TOC recognizes getting a homeless person from the streets to a home is not a linear process. It can involve employment training; mental health counseling; healthcare; childcare; substance abuse counseling; and much more.

Being able to rapidly adjust a journey across all of your channels will become a hallmark of high-performing organizations. Whether it is a journey from Member to Major Donor or the journey from alcoholism to sobriety, you will not reach your destination unless you focus on the journey.

There is a growing list of software vendors offering journey mapping tools. Smaply (https://www.smaply.com/) offers a 14-day free trial and it is not too expensive if you get hooked on it. Just going through this exercise with your team to try to map your donor or mission beneficiary experience will yield valuable insights into how connected your organization is with the experiences your constituency is having with your organization every day.

Personality Analytics

Most of us have taken a personality test either to get a job or out of curiosity. You may be an "ENFP" which stands for Extraverted, Intuitive, Feeling, and Perceiving or an "ISTJ" which stands for Introverted, Sensing, Thinking, and Judging. If you know your letters, you have likely taken the Myers-Briggs Type Indicator® test.

My personal favorite is from a book, *StrengthsFinder 2.0*, written by Tom Rath. The book includes a code to access a test focused on your strengths rather than your weaknesses. Of course, your strengths are a big window into your weaknesses, so you still get all the benefits of a traditional personality test while keeping the focus on working with what you have rather than what you need to fix.

If we could get our constituents to take these tests it would really help us tailor our communications and touch points. That is clearly not an option, so we have been left to figure it out over time as we have more interactions. It is not surprising some of the best fundraisers, and counselors, are people who are able to intelligently listen. For them, every word you write or speak is piece of a personality puzzle. Knowing what to ask to get the next piece is the true art of building relationships based on who a person truly is and what they need from you.

This has worked very well for major gift fundraising, and for one-on-one work with mission beneficiaries, as long as the staff member in the organization has 'a trained ear. The problem comes early in a relationship when the relationship may be based on email or direct mail copy. Your communications may be resonating with certain constituents, but falling flat with others.

Enter cognitive personality testing: the ability to ascertain a person's personality traits from text they have authored. The science behind this is based on the psychology of language. You are using the words a person uses to determine their personality type, values, and needs.

Where do you get authored content?

- Emails
- Letters
- Social media posts, comments, and tweets
- Articles
- Audio transcriptions

With IBM's Personality Insights™, you can extract 52 aspects of a person's personality. From Extroversion to the need for Stability you are compared to the sample population they used to train the service. This means for a particular personality element you might have a 65% score, meaning you exhibit the trait greater than 64% and less than 34% of the sample population.

What does this all mean for our sector? Let's break it down into different parts of the organization:

Communications: You can study the personalities of people who have responded, or not, to different channels of communications to help you better use those channels and match your copy to personalities more effectively (also use Tone Analytics).

Fundraising/Volunteering: Match field officers to prospective donors based on the best personality fit. This can be especially useful with high-potential donors who have very distinct personalities. Augment your current HR processes to ensure fundraising team members are a good fit. Bring personality science into event seating to make it more enjoyable for the attendees. Find people who are the best fit for particular volunteer opportunities.

Mission: Augment your current HR processes to find the best candidates for positions in your mission operations. This can also help you get people in the right seat on your organizational bus. Use to help better understand the personalities of your mission beneficiaries where applicable. Find people who match up best with non-fundraising volunteer activities.

Call Centers/Chatbots: Understand how the personality of your callers and call takers impact their performance. When using chatbots, route calls to specific live-operators based on personality matches. Combine personality with tone analysis to gain an even more accurate picture of how calls are going

It is understandable to worry these insights into personality might cloud your judgment. Remember, this is augmenting, not replacing, a person who is actively managing a relationship. What it is replacing is random communications and fundraising appeals.

Another reality is we bring a great deal of Confirmation Bias when we are judging someone's personality. Because of this, Cognitive Computing may actually offer a more level playing field for qualified job applicants who do not look like us or come from the same background.

With Personality Analytics organizations can also deepen and strengthen relationships in real-time as their communications with constituents grow. All organizations struggle with bringing a newly-acquired donor to the next level or even keeping them at the current level. You can see this in the low retention rates and flat fundraising results.

We cannot afford to wait until an organization's most precious resource, people, can be allocated. It is one of the ironies of this technology which so many people fear will replace humans – it actually humanizes the non-human aspects of our operations.

Behavioral Analytics

Do you want to be Sears® or Amazon? As I'm writing this, the company that now owns Sears is talking seriously about closing down for good. How could the company that invented large-scale retail, and the catalog business, be in danger of extinction?

The demise of Sears will be studied for years to come, but it boils down to an inability to change with its customers. Sears offered departments within its stores, mimicking Toys"R"Us; Home Depot; Best Buy; and every other niche store, yet they wouldn't accept shoppers preferred specialty stores selling toys; hardware; and electronics. The data was there, the insights were there, but the executives put their heads in the sand and chose not to see it.

What is even more maddening is Sears smartly branded their appliances Kenmore® and tools Craftsman®, yet could not make the leap to create stores around them.

And then came online retailing. Why did Sears not see this potential? It was their chance to put their catalog on the web, and, instead of Amazon, there would be Sears.com (it exists, but when was the last you went there?). I'm sure as they looked at the customers in their stores, who were increasingly older, they came to the conclusion people were never going to put their credit cards on the internet. What I will never understand is why they didn't see that people would ultimately want to have a wide choice of items and would gladly wait to have it arrive at their home just as Sears had taught us to do with their catalog.

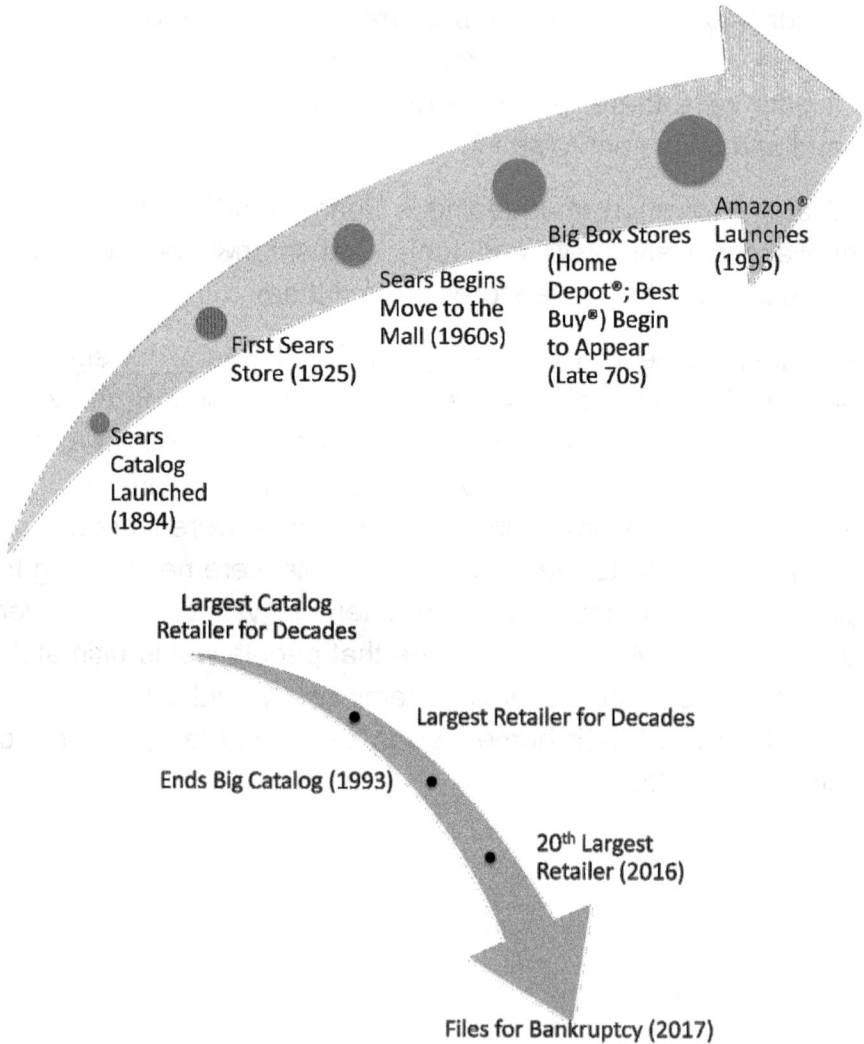

Sears Catalog Launched (1894)

First Sears Store (1925)

Sears Begins Move to the Mall (1960s)

Big Box Stores (Home Depot®; Best Buy®) Begin to Appear (Late 70s)

Amazon® Launches (1995)

Largest Catalog Retailer for Decades

Largest Retailer for Decades

Ends Big Catalog (1993)

20th Largest Retailer (2016)

Files for Bankruptcy (2017)

This is a cautionary tale for every organization thinking they *know* their customers. To move forward effectively in a world where consumers are continually changing their behavior, it is imperative to have a strong Behavioral Analytics program.

So, what is Behavioral Analytics? It is the study of how people behave as they interact with your organization, and as they behave outside of your organization. Think of donors who support your organization; people who support other organizations, but not yours; and people who support both.

You have deep information on their giving and interactions with you. You will want to append demographic, economic, interest, and philanthropic data. You will also want to find published studies about donors in general as well as specific types of donors. This will give you context to review your data. In the case of Sears they would have seen that while they had the older shopper locked down, they were not picking up the new generation (yes, the Boomers were once the new generation). They also would have seen the move to specialized stores and subsequently to online shopping.

Cluster Analytics is a key component of Behavioral Analytics. While you do ultimately want to understand each individual constituent, you will want to work with clusters as you design experiences, and create new journeys, to work with them.

- One of the early wins from Behavioral Analytics is realizing people can act differently when they interact with your organization than they do with others. A number of years ago I read a study of Target® shoppers; Walmart shoppers; and Target and Walmart shoppers. A couple of items jumped out: first, with people who shop at both stores, certain types of shoppers spend very different amounts of money; and second, people who only shop at Target or Walmart (not both) also shop at other stores which are very different from each other.
  - Professional women were spending a lot more money at Target for clothes than at Walmart.
  - People who shop at Target are likely to have recently shopped at Macy's® while people who shop at Walmart are likely to shop at Dollar General®.

Wal-Mart used this insight to create stores designed to appeal to professional women. Target focuses a lot of their product buying on having items that look like what you would find at a Macy's, but less expensive.

How could you do a study like this? Use a data matching service that offers philanthropic giving data, and see if there are organizations to which a large number of your constituents also give. Pick one of the organizations and create the following constituent groups:

- Constituents who only give to your organization.
- Constituents who give to the organization you have chosen.
- Constituents who only give to the organization you have chosen.
    - This group can be members who have not become donors; non-donor alumni/ae; or patrons who are not donors.

Now look at giving; engagement scores; and other data to determine if there are any patterns among these three types of constituents.

Bring NLP to the table by analyzing your selected competing organization's marketing and fundraising material to see what they are doing to attract your constituents. There is nothing wrong with borrowing from a good idea. Odds are your "competitor" has taken a look (albeit not with NLP) at what your organization is doing.

Another aspect of Behavioral Analytics is people will act differently within different channels. This is important because if you look at your overall success you can miss that people dislike a channel, yet give in another, resulting in the false sense that all is well.

It could be they are just not responsive to email (and maybe even consider your emails spam), yet they love direct mail. Try cutting back on your emails, and see if that increases your giving.

This brings us to the problem of making choices. The easiest one is to just do everything with everyone. Have an email – send it to all. Direct mail campaign goes to all people with good addresses.

This is your Sears or Amazon moment. This is when all of your NLP; Journey Analytics; Cluster Analytics; and everything else you have done to glean insights into what people are interested in about your organization springs into action.

Are you going to be Sears with everything spread out, hoping at some point people will see what they like and buy it? Or will you narrow the ways you communicate, and what you communicate about?

When you go to Amazon, every product imaginable (for purchase) is there, but what you see on your page one is based on their Behavioral Analytics. They know there are things not on that page you may well want more than what is, but they trust that they will do better by narrowing your choices.

Deciding to be purposeful with your communications will require not just marketing and fundraising skills, but also political acumen. Whether it is the person in charge of a communication channel or a person who manages an area of your mission, they are not going to necessarily understand why you are making these choices.

This is where how you present your insights will really matter. We cover that in detail in the chapter Visualization – Tell Your Data's Story, but for this particular challenge I would be sure to show each channel, and each area of interest, what they are getting out of this as well as the gain for the overall organization.

Done right, this is not a zero-sum game where one channel or area has to lose for others to gain. Take advantage of the tough truth at every organization – constituents are not giving anywhere near their giving potential. I have worked with many of the Philanthropic 400 and seen in rich detail their fundraising results, and none of them come close to maximizing the giving potential of their constituencies.

This is not a knock on those organizations; it is just a reality that our current approach to engagement and fundraising focuses either on mass-communication or very high touch relationship management. That leaves a huge number of constituents touched only by poorly targeted communications that are sent in the hope something sparks their interest.

A trend I am seeing in marketing is a move away from vertical products designed for our field. I call these the Microsoft Office® products, seeming to be fully integrated with your donor management systems. The phrase "one-throat to choke" might ring a bell here.

Underutilization of true omni-channel marketing masked the limitations of these systems. In fact, even when this was implemented it was often out-sourced. Now organizations have people on staff who can implement dynamic multi-channel campaigns, but not with the traditional tools.

Companies like Marketo® are gaining traction, and this is expanding the vision of what can be done. One of the first areas to see the impact of best-of-breed technology was admissions. Today you can see email, web, social media, and direct mail coordinated to match the reality of today's consumer. We no longer think of a person being an *on-line donor* or *direct-mail responder*, but instead recognize people want to move seamlessly across all channels.

While this is exciting, it also has exposed challenges with current data platforms. Multi-Channel systems are hungry for insights. While they will create their own with interactions, they will not have all of the data from your other touch points. One of the top ROIs on your Insight Reservoir will be supporting these platforms.

On the mission side of your organization, there is the opportunity to dramatically improve service delivery. Nick Intintolo, an Executive with IBM, has used Big Data and Cognitive Computing concepts to develop Human Outcome Analytics to help human service organizations use current and historical data to better predict outcomes. Improving predictive accuracy can be the difference between life and death in areas such as child welfare services.

As you explore behavioral analytics, keep in mind Sears did not fail because people stopped shopping. They failed because they lost touch with their customers and were unwilling to provide those customers the experiences they wanted. If you have an aging file, you might want to hang a picture of a closed Sears in the conference room. Sadly, there will be plenty of them all too soon. The point is, make sure your organization is adapting to change with the needs of your constituency and the experiences they wish to have with your organization.

Social Analytics

At one point, I owned every Who's Who® in print. Only a data geek would be willing to admit his home was once filled with volumes he had purchased from musty used books stores. When compiling profiles on philanthropists I would start here (actually I started with the Biographical Master Index® which referenced the books). Endorphins were released whenever I found a donor because it provided a map to finding everything else I needed, from family to business to philanthropy.

With this in mind, you can understand how excited I was when LinkedIn® came along and suddenly instead of outdated short biographies, I had access to hundreds of millions of resumes. More than one information professional was not as enthused, fearing people may lie on their profile, but I pointed out Who's Who sends a questionnaire for people to fill out themselves, which sounds a lot like LinkedIn.

My wealth screening company, Prospect Information Network (P!N), was originally an idea for a social network to serve our sector. The clue is the word "Network." I was inspired in part by the book, *Net Gain: Expanding Markets Through Virtual Communities*, by John Hagel, III, and Arthur G. Armstrong. They really nailed how to build and monetize a social network. Fortunately, they were also very frank about the challenges to make it work.

It was 1997, and I came to the conclusion the World Wide Web was too immature to base a business on, and especially one that expected the free exchange of personal information between people. I have no regrets in going back to my roots and making P!N a wealth screening company because I am 99.9% sure a social network (or virtual community as it was known then) would have failed at that time. The 0.1% is a nod to my entrepreneurial optimism.

The turning point for social media was in 2005 as Facebook® changed social networks from backstage with the band and dorm rooms (MySpace®) to communities, and suddenly a rich source of personal information became available. Here again, leaders in our sector at first were not too enthusiastic because they assumed only young people were members, not their donors who are typically middle-aged and up. I don't hear that anymore as Facebook has grayed.

Today, of course, we have Twitter®; Instagram®; Pinterest®; and a host of specialized communities covering different niche markets. Each of these social networks are stronger with different demographics, but together create an incredible resource of insights on your constituents.

Social media is producing so much information, it sometimes is thought of as Big Data rather than just one of the sources for it. Here are some things people do every minute:

- Facebook® has over 4 million users like posts.
- Instagram® posts have 1.7 million likes.
- Twitter® users tweet 350K times.

  http://wersm.com/how-much-data-is-generated-every-minute-on-social-media/

That is a lot of activity, which could easily overwhelm even the most robust Big Data platform. The key is capturing activity relevant to your organization and then connecting the activity to your constituency.

Step one is identifying all of the social media networks managed by your organization. Cast your net wide to ensure you capture all of them. College and universities have sites for schools; sports teams; performing arts centers; and other connected organizations. NGOs may have a primary site, but also sites related to particular locations, programs and/or events. A hospital system may have sites for each facility as well as auxiliaries and events.

Once you have a complete list, determine who manages each network. Hopefully, they are centrally managed. If not, you may need senior leadership to break through the politics for you to gain access.

What you will need to bring this data into the Insight Reservoir is to be made an administrator for the social accounts. This will give you the right to mine the data. You can still find value in social media without this access, but it is greatly diminished. It's worth the political battle.

Your next step is to match your constituents to all of the networks. This is done using paid services and is based primarily on emails. Finally, you will set-up a monitoring system to bring all of the posts; likes; shares; and tweets into the Insight Reservoir.

This is when the analytics begins. One of the first insights will be constituent engagement across your networks. Look for people engaged in more than one network, and see how that is correlating to giving and retention. You will also be able to infer interest for the networks dedicated to particular sports, schools, projects, and programs.

You will use NLP to extract insights from tweets, comments, and posts. Visual Recognition will turn pictures into windows into people's connection with your mission.

One of our projects at NewSci connected 45 social networks to over 100,000 alums. Being able to see not only the activity at the university's primary social pages but also at the individual schools plus all of the athletic teams provided actionable insights about engagement levels and interests.

Another reason to have a strong Social Analytics program is the increasing use of social media to drive fundraising campaigns like #GivingTuesday. Social media is now so woven into our lives we no longer see it as a side channel of communication or as something we just do to communicate with friends. All kinds of commerce is happening on social media, and helping others is very much a part of the experience.

All of this sounds incredible, but what about privacy?

Even as an administrator, your access to social profiles will still be governed by user privacy settings, and also by the terms and conditions of the social network. Constituents who engage with your social networks are not surprised you are seeing their interactions. In fact, not acknowledging what is being said is a modern version of not answering your phone or responding to a letter.

Social analytics is a vital source for a successful Insight Reservoir. It will provide a real-time view into your constituency, and support many of your other analytic activities.

Spatial-Temporal Analytics – The Where & When of Analysis

When you find yourself talking about space and time you are either at a scientific convention or a party where you should have asked what was in the brownies. Add a book about Cognitive Computing to the list.

When it comes to data science, knowing where your data originated (spatial) and when your data was created (time) is essential to gaining accurate insights. This is not true for all questions you will ask your data, but when it matters it matters a lot.

For example, let's look at your last fundraising campaign. You most likely recorded the date of each gift, and the location of the constituent who supported you. Sounds like time and space, right? It depends on what you are trying to analyze.

The date of the gift is the date you received the gift unless it was on-line. This means the person mailed the check on a different date. Depending on how far they live from your headquarters, it could have been a week before you received it. As for location, online giving can throw you off. Perhaps they received your email while on the road, and made the gift far away from their home.

Why would any of this matter in your analysis? It would if you wanted to know what impacts your fundraising results. One of those impacts could be weather. Thanks to The Weather Channel's data (now owned by IBM), you can have the weather for any location going back years.

Let's say you did a mailing that dropped on March 1st. You can estimate the delivery date within a day or two across the country for each location, and look at the weather on those days. Did any locations underperform that had bad weather? I know here in Florida when it is hurricane season, our focus shifts from our mailbox to our power and roof. Weather has been used very effectively by museums and zoos to better understand when to expect big crowds, and when to staff-down.

Schools can use Spatial-Temporal Analytics to study the movement of students on their campuses. The inputs could be connections to the school's Wi-Fi network and/or the school's student app. Before you get worried about privacy, schools can easily make this kind of study not involve student names. What they need to know is where people are going and when they are going there. This can help with safety by showing areas that need better lighting because students are going there in the evening.

Embedding time and location data with economic metrics such as unemployment enables you to see whether your results are because of your message or the reality your constituents are living in. If you are a national or international organization, this becomes even more important as it is easy to obscure local success and failure under the weight of the total results.

As we were recovering from the Great Recession, I would use the Coincident Index, created by the Federal Reserve Bank of Philadelphia, to illustrate to clients how different parts of the country were doing better or worse at the time. This is one of the few economic indicators designed to show you how the economy is doing now rather than in the future. Instead of being a leading or lagging indicator, it is meant to coincide with the economy. At the State level you can put the monthly data on top of your results, and see whether your results are rising and falling with the economy or against it.

To give you an idea of how compelling this is, in December 2008, only Louisiana, North Dakota, and Wyoming were showing positive growth. In July of 2017, 41 States were showing positive growth. You can find the current and historical data at https://www.philadelphiafed.org/research-and-data/regional-economy/indexes/coincident/maps.

Transportation is one of the biggest users of Spatial-Temporal Analytics. After all, it is worth a lot of money to find efficiencies in getting people and products from point A to B. In your organization, it might be how to better plan the travel of your major gift officers to increase the number of highly rated prospects they can see each year. On the mission side, it might involve delivering more services to more people in the same amount of time.

Take a look at your face-to-face fundraising meetings and group them by location, and time. How many trips did you take to a particular city, and how many people were you able to see on each trip? Now, look at your prospect pool and see if through better planning you could increase the number of visits 10-20-30%. What would that mean to your results?

This type of analysis will also help you better assess the value of travel. Too often we focus only on the trip involving the ask, when in fact there may have been one or more cultivation visits. What about stewardship? Many a major gift officer has told me they left their last organization in part because they were so bogged down in post-gift activities they couldn't see new prospects.

What is the optimal mix of cultivation, ask, and stewardship meetings? Odds are that will depend on who is doing the traveling: A new gift officer or the person who was your best closer the last campaign?

As Cognitive Computing becomes more embedded in CRM systems such as Salesforce® and Microsoft CRM, this type of analysis will become standard. You can imagine your CRM records being linked to flight schedules to maximum not only time, but also lower cost.

Don't forget all the other data, such as gift capacity, you can layer in to see clearly where, and how, you should be spending your most valuable resource - time. Those stewardship calls that seem like a drag on your results may actually be critical to your long-term success.

You will discover as you bring time and space into your analysis the insights and the conversations deepen. When a report, spreadsheet, or chart was prepared will have far less meaning than when and where the data contained in it happened.

Visualization – Making Your Insights Come Alive!

All of the insights you glean from your work will be for nothing (or at least worth a lot less) if you are unable to communicate them to the people who can act on them. This is where visualization comes in. Companies like Tableau®, Qlik®, and Domo® have taken us from Excel charts and PowerPoint® slides to an interactive experience where you can turn your insights into a story.

While I enjoy the current platforms, I believe we are just at the beginning of what will come as we seek to fully comprehend, and act on, the deep insights from our data. It will not be long before it will be commonplace to interact with data using virtual reality. You will be able to explore clusters and take the journey your constituents take in a three-dimensional world.

The idea of dashboards will be as archaic as floppy disks and modems. Insight assistants will be in the conference room, operating room, and homeless shelter or wherever you are. With a simple question, you will be provided the latest insights.

This is already being tested at Rensselaer Polytechnic Institute where a room has been set-up to study how people interact with Cognitive Assistants. Will a doctor accept the advice of a computer? Will you take into consideration an idea put forth in a meeting by your version of Watson? You might think not, but keep in mind it was not that long ago we all were fine with navigating a new city based on a printed map. It was also only a couple of decades ago we didn't have the web to give us answers.

Telling a story with data takes some adjustment. At first, you will just present a prettier version of a spreadsheet. I suggest you don't start with the data, but rather with the idea you are trying to present. This approach frees you from thinking about how you are going to create the visuals.

One idea is to think of a three-act play: Introduce your idea; take them through a journey; and have an ending that leaves them smiling or crying.

Act 1) Lack of engagement with our younger donors.

Act 2) The journey our younger donors take through our fundraising process and the poor results.

Act 3) Outcomes if we change our approach.

You might start with a simple bar chart of giving for constituents age 20–35 over the last 3 years. You can then show the same group, but in clusters with explanations of who is in each one. Remember, with Cluster Analytics you are able to create clusters within segments such as age. You now have Act 1.

In the second act you can show the journey all donors go through in your organization with total giving by age. Now show the same for the 20-35 age group clusters. This is where you can highlight differences between how they react to certain touch points vs other generations. Maybe they like crowdfunding, but are not responding as well to telemarketing and direct mail.

For the final act, you can show the best performing clusters in terms of total giving and average and/or median gifts. Then show them what would happen if you were able to get 25% of the underperforming clusters to perform like the highest one, and be ready for people to ask you to show higher percentages.

Take your bows, and begin the conversation of how you make the changes to the journey to get the results you are seeking.

Initially, your visualizations will likely be just slightly improved versions of what you have done before. This is understandable, as we have been conditioned by spreadsheets to accept bar charts and pie charts as the norm.

To break out of this visual rut, download sample visualizations from vendors. It will not likely have anything to do with your organization, but it will spark your imagination.

Instead of a pie chart try a heat map. Instead of a bar chart try a bubble chart.

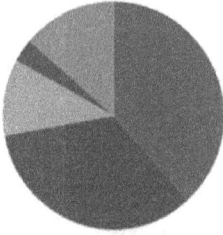

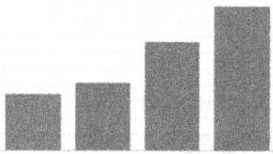

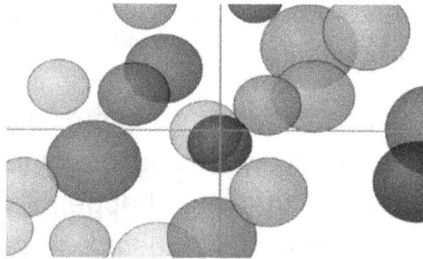

Remember you now have text, so use it! This means not just word clouds, but also more imaginative ways to show how your text is connecting to outcomes. Create topic clouds where a central idea is in the center, and the related words are around it. This is especially useful when trying to determine if there is positive or negative sentiment about a topic or issue.

Another consideration is how you group your visualizations into dashboards. They need to reinforce each other and also add depth to the analysis. Remember your audience is not necessarily going to get it or see it like you do. It will take some trial and error to ensure your visualizations resonate with your audience.

Finally, your visualizations and dashboards will be put together into a story. As I mentioned earlier, treating it like a three-act play will make it feel more natural to your audience.

Be careful not to jump around between topics. If you have a lot of topics to cover, be sure to complete your story for each before moving on. The third act may be showing how all the topics come together.

As visualization tools become the norm for delivering insights to end-users, we are also starting to see a blending of narrative and graphics. This is important, especially when creators of insights are not able to personally provide context. The Tableau Wordsmith® partnership is an example of how interactive insights will be delivered.

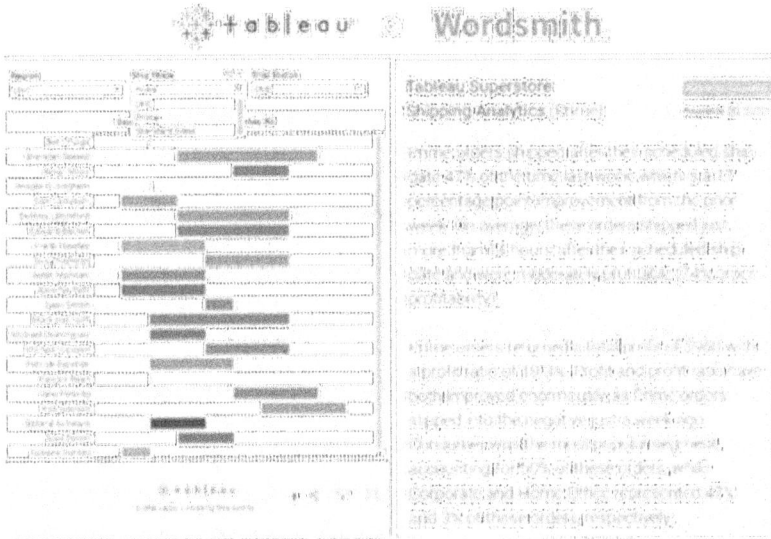

You can see an organization's annual report delivered in this format, allowing supporters to journey into your mission through visualizations, while also being guided by the narrative.

Cognitive tools are being introduced constantly, and the flow is only going to increase. We are at the beginning of this age. In the early days of relational databases, the reporting and query tools were poor or non-existent. This is where we are today with Cognitive Computing. They are better than the past, but still primitive.

Where I see this going is to a time where insight walls will be standard parts of conference rooms. Insight walls will be rich multi-dimensional graphic versions of today's whiteboards. Our voice, eyes, and pointing of our fingers will power the experience.

Conference tables will be replaced with holographic images of what we are discussing. The images will change with the conversation, not with a click of a device. These Insight Rooms will be designed to enable us to experience our data rather than passively view it. It will encourage creativity, as we are free to ask questions as they come to us rather than having to prepare all of the reports prior to the meeting.

This may seem like science fiction, but so do all things when they begin to be first imagined.

08

Crowd Science – Your Data Science Volunteers

> "Only a life lived for others is worth living."
>
> —Albert Einstein

Finding data scientists is a major challenge for large successful companies, so it is safe to assume NGOs are going to find it hard to attract and retain top talent. Colleges and universities are finding it difficult to attract and retain top faculty in data science. Does this mean organizations will have to sit on the sidelines while their for-profit brethren enjoy the benefits of Cognitive Computing? Absolutely not!

Your organization may not have the money data scientists have come to expect, but you do have something they often love even more – data. Machine Learning has an insatiable appetite for data, and the more the better.

Kaggle (www.Kaggle.com) is an example of the growing data science crowdsourcing community. You can create a competition around your data, and the data scientists in the community begin their work. There are cash prizes, and your organization might want to put some money in the pot, but other competitions involve just receiving swag.

Here are some current and past competitions:

UPenn and Mayo Clinic's Seizure Detection Challenge

Detect seizures in intracranial EEG recordings

$8,000 in prize money

200 teams participating

Passenger Screening Algorithm Challenge

Improve the accuracy of the Department of Homeland Security's threat recognition algorithms

$1,500,000 in prize money

212 teams participated

The Nature Conservancy Fisheries Monitoring

Can you detect and classify species of fish?

$150,000 in prize money

2,293 teams participated

The Marinexplore and Cornell University Whale Detection Challenge

Create an algorithm to detect North Atlantic right whale calls from audio recordings, prevent collisions with shipping traffic

$10,000 in prize money

245 teams participated

Melbourne University AES/MathWorks/NIH Seizure Prediction

Predict seizures in long-term human intracranial EEG recordings

$20,000 in Prize Money

478 teams participated

Not all of the competitions are mission-critical. You can hone your skills trying to predict who will survive the Titanic - https://www.kaggle.com/c/titanic.

DataKind (http://www.datakind.org) is an NGO, and they are focused on providing data science for the good of humanity. Here is their mission statement:

> We are living inside a data revolution that is transforming the way we understand and interact with each other and the world - and it has only just begun. Every field is now having its "data moment," giving mission-driven organizations brand new opportunities to harness data to advance their work. From poverty alleviation to healthcare access to improved education, data science has the potential to move the needle on

seemingly insurmountable issues, but only if there is close collaboration between data science and social sector experts.

Jake Porway, the Founder and Executive Director of DataKind, has done an amazing job bringing together talented data scientists; funders; NGOs; and subject matter experts. Here are some of the projects DataKind has worked on:

Forecasting Water Demand in California When Every Drop Counts

> Develop a "proof of concept" water demand forecasting model using flow data at the micro zone level for potential future scaling to other retailers in the California Data Collaborative, a unique water manager-led public private partnership that brings together utilities across the state to leverage data to help water managers ensure reliability.

Advancing Financial Inclusion in Senegal Using Predictive Modeling

> Using Microcred's loan application data and internal loan status information, build statistical models to help predict customer default and better inform decision making about lending to make it more efficient and inclusive.

> Identify whether loan application data and factors such as past repayment behavior or late payments during the early stages of a loan cycle could predict default.

Using Open Data to Prevent Home Fire Deaths and Injuries

Prepare and anonymize never-before-released American Red Cross domestic disaster data and open data to gain insights

Explore the data descriptively with statistics and visually with maps and graphics to understand data quality and identify patterns

Prototype a model that predicts and maps which counties American Red Cross should target for its next smoke alarm installation campaigns

The Science for Social Good Fellowship at IBM Research is an opportunity for NGOs to have both the resources of IBM, and the time of researchers, to work on their challenges:

The Science for Social Good Fellowship at IBM Research is an opportunity for undergraduate and graduate students as well as postdoctoral scholars to develop their skills and prototype new technologies that benefit humanity. Mentored by leading IBM Research scientists and engineers at the T.J. Watson Research Center in Yorktown Heights, NY (north of New York City), fellows use data science, machine learning, AI, analytics, operations research, statistics, design and mobile computing methods to complete projects with social impact. Working closely with non-governmental organizations, social enterprises, government agencies, and other mission-driven partners, fellows take on real-world problems in health, energy, environment, education, international development, equality, justice, and more.

http://researcher.watson.ibm.com/researcher/view_group.php?id=7267

Here are some recent projects:

Open Discovery Platform for a Multiple Sclerosis Cure

Partner Organization: Accelerated Cure Project for Multiple Sclerosis

Motivation: Multiple sclerosis (MS) is a disabling disease of the central nervous system that disrupts the flow of information within the brain, and between the brain and body. The cause of MS is still unknown: scientists believe the disease is triggered by as-yet-unidentified environmental factor in a person who is genetically predisposed to respond. The progress, severity and specific symptoms of MS in any one person cannot yet be predicted. Analyzing heterogeneous data from clinical trials, clinical blood reports, gene-wide association studies and many other sources can lead to the discovery of factors associated with the disease. However, a single actor cannot hope to find such factors working alone; a platform for anyone in the world to collaboratively determine hypotheses to test and patterns to mine, with the assistance of cognitive technologies is a must.

Project Outcome: We developed a unique set of cognitive capabilities that understand the source code of a data analysis without any intervention by the user. This approach allows one to compare analyses based on an ontology, enabling a recommendation of similar and complementary analyses, visualization of the space of analyses, and so on, thereby accelerating discovery and knowledge sharing. Preliminary data analyses on factors that trigger multiple sclerosis, conducted using a large, heterogeneous MS database, have been analyzed through the cognitive capabilities.

Hunting Zika Virus with Machine Learning

Partner Organization: Cary Institute of Ecosystem Studies

Motivation: The flaviviruses are some of the most widespread viruses known. West Nile, Yellow Fever, and Dengue are a few of the best-documented examples in the Americas. The recent emergence of Zika virus outside of Africa has reiterated the need to discover what suites of correlated features of mosquitos and wild mammals combine to describe the most competent vectors and identify which wild primate species should be targeted for viral surveillance and management.

Project Outcome: We used physiological, behavioral, range, and social structure data of mammal species to develop a Bayesian predictive model of their status as reservoirs for different zoonotic diseases. The approach was trained on reservoir species from Africa to predict reservoir species in the Americas. The results will guide focal testing by disease ecologists in the field.

What Works in Global Development?

Partner Organization: Clinton Global Initiative (CGI)

Motivation: CGI facilitates innovative solutions to the world's most pressing problems through Commitments to Action that its members make and report measurable progress on. Best practices and lessons learned can be analyzed, captured, and disseminated from the corpus of 3,500 Commitments collected over 12 years. What best practices can be drawn from the Commitments that are relevant to others working in this space? How can work in this space be successfully scaled and replicated? What types of partnerships exist within these Commitments? What types of partnerships are most successful? What are key lessons learned as it relates to cross-sectoral partnerships? What common challenges exist across these Commitments?

Project Outcome: We developed a recommendation engine for Commitments by way of meta-analysis using natural language processing of Commitments and progress reports, along with network analysis of the members and partners involved. The recommendation engine can be used to guide new philanthropic investments and partnerships. We also developed a visualization that helps convey the advantages of bringing together different types of organizations to conduct social good projects.

Disseminate the Best Treatment for Diarrhea

Partner Organization: Clinton Health Access Initiative (CHAI), Inc.

Motivation: Diarrhea is the second leading killer of children under 5, responsible for more than 700,000 deaths globally each year. Zinc and oral rehydration salts (ORS) can prevent over 90 percent of diarrhea-related deaths and cost less than 50 cents per child; yet few children in need are receiving treatment. At the start of the program in 2011, an estimated 32 percent of children with diarrhea received ORS and less than one percent received the full recommended treatment. Instead, the majority of children would receive suboptimal products like antibiotics and anti-diarrheals or nothing at all. In Nigeria, CHAI worked with the National Association of Proprietary Patent Medicine Dealers (NAPPMED), the trade union for over-the-counter medicine vendors, to train representatives to visit other medicine shops and teach their peers about ORS and zinc, with the goal of increasing their stocking of the treatments.

Project Outcome: We performed statistical analysis on CHAI's program and found medicine shop owners who were visited by a representative were more likely to have correct knowledge of the best diarrhea treatments and were more likely to stock ORS and zinc. The results will help CHAI refine their programs around the world.

Real-Time Understanding of Humanitarian Crises

Partner Organization: ACAPS (Assessment Capacities Project)

Motivation: Over the last decade there has been notable improvement in the collection and dissemination of humanitarian information used to analyze the nature and magnitude for major crises around the world, and suggest better targeted response priorities. There has been healthy growth in the recognition of assessment and situation analysis as vital to humanitarian program design and monitoring.

Through the last few years the systematic use of secondary data to inform and provide context for emergency programming has become a norm. At the same time, this has led to the emergence of analysis units and experts across thirty agencies and NGOs doing similar labor-intensive humanitarian assessment. These activities can be automated and improved using machine learning and information retrieval techniques.

Project Outcome: We developed an end-to-end system accessible through an API that performs focused web crawling to bring back articles only relevant to humanitarian crises, classifies them by type of disaster, and provides a faceted search interface to access the results using Watson technologies. The system improves the quality and depth of humanitarian analysis using data science techniques to take advantage of the increased volume and variety of available information.

The Data Science for Social Good Fellowship at the University of Chicago (https://dssg.uchicago.edu/) trains data scientists by working on data science projects with social impact.

The Data Science for Social Good Fellowship is a University of Chicago summer program to train aspiring data scientists to work on data mining, machine learning, Big Data, and data science projects with social impact. Working closely with governments and nonprofits, fellows take on **real-world problems** in education, health, energy, public safety, transportation, economic development, international development, and more.

For three months in Chicago they learn, hone, and apply their data science, analytical, and coding skills, collaborate in a fast-paced atmosphere, and learn from mentors coming from industry and academia.

Here are some of their projects:

As part of the **2015 White House Police Data Initiative**, DSSG partnered with multiple police departments, including the Charlotte-Mecklenburg Police Department, to apply data analysis to identify which factors should be used in early warning systems to flag at-risk officers before an adverse interaction occurs. Using anonymized police data, as well as contextual data about local crime and demographics, this model detects the factors most indicative of future problems, so that departments can provide additional support to their officers. In 2016, we partnered with additional police departments, including the Metro Nashville Police Department, to test and expand this work in new municipalities, improving both the overall model and local performance. Like our work in 2015, we used anonymized police data and contextual data about local crime and demographics to detect the factors most indicative of future issues, so that departments can provide additional support to their officers.

Both departments continue to work with the Center for Data Science and Public Policy on implementing the new EIS. The Nashville team gave their department a list of the highest-risk officers according to our model, which MNPD subsequently used to send letters to the officers and their supervisors informing them of the results and specific risk factors that led to their score.

We're now helping them integrate the EIS into their existing IT system, so that it will continuously update with new data. https://dssg.uchicago.edu/project/early-intervention-system-for-adverse-police-interactions/

Lead paint and leaded gasoline were banned in the United States in the 1970s because of the enormous public health dangers lead poses. In the decades since, it has become clear that even small amounts of exposure to lead during childhood can cause behavior or attention problems, learning difficulties, speech and language problems, reduced IQ and failure at school.

Over the past several decades, Chicago has made great strides in preventing exposure to lead. Even with this progress, there is more work to be done. In 2013, it is estimated that almost 9000 children in Chicago had been exposed to levels of lead that the CDC classifies as dangerous, and that most of this exposure happened in the home.

In 2014, DSSG partnered with the **Chicago Department of Public Health** to help find the homes that are most likely to still contain lead-based paint hazards. By building statistical models that predict exposure based on evidence such as the age of a house, the history of children's exposure at that address, and economic conditions of the neighborhood, CDPH and their partners can link high-risk children and pregnant women to inspection and lead-based paint mitigation funding before any harm is done. This integrated and innovative system will ensure resources are used most efficiently, and ultimately will mean healthier Chicago children.

https://dssg.uchicago.edu/project/predictive-analytics-to-prevent-lead-poisoning-in-children/

Electronic medical records (EMR) promise to transform our understanding of patients' ailments and improve their care. **NorthShore University HealthSystem** in suburban Chicago has been a national leader in the implementation of EMR systems for the past decade. It is the first healthcare provider to be awarded the highest level of EMR deployment for both inpatient and outpatient care. This remarkable effort has generated much anonymized data available for innovative analytics research.

We worked with NorthShore scientists to tackle:

1. Childhood obesity: Growth charts are percentile curves that illustrate how kids' height and weight change during childhood. Surprisingly, these growth curves are one-size-fits-all: there's just one version for each sex. Our team built personalized growth curves, allowing doctors to detect childhood obesity earlier and enabling them to intervene early.

2. Code blue: when a patient goes into cardiac arrest, medical staff issue a code blue alert. Staff stop what they're doing and attend to the patient and yet less than 80% of victims survive. We are working on predicting these medical crises before they happen, allowing doctors and nurses to intervene before patients have cardiac arrest.

https://dssg.uchicago.edu/project/using-electronic-medical-record-data-to-predict-better-health/

The maternal deaths in Mexico from pregnancy, childbirth or postpartum complications have decreased from 89 deaths per 100,000 live births in 1990 to 43 in 2011. Despite this improvement, the rate of decline has significantly slowed and Mexico was not on track to achieve its Millennium Development Goal of reducing maternal mortality 75% by 2015.

Working with **Mexico's Office of the President**, our goal was to identify factors contributing to maternal mortality and determine what could be done to reduce it. In contrast to previous work, we analyzed trends at a more granular level. While our initial work focused on municipalities and localities, our intention in 2014 was to develop individual-level models of risk using all available data. https://dssg.uchicago.edu/project/reducing-maternal-mortality-rates-in-mexico/

Our team is working with the **City of San Jose's Code Enforcement** office to identify the multiple housing properties at highest risk of violations for more immediate and frequent inspections. With more than 4,500 multiple housing properties in San Jose — many of which comprise multiple buildings and hundreds of units — it is not possible for the city to inspect every unit every year. Some properties are less well-maintained and require more attention from inspectors than others, making prioritization of properties an important aspect of identifying violations in a timely fashion.

San Jose recently instituted a tiered approach to prioritizing inspections, inspecting riskier properties more frequently and thoroughly. The tier system provides an incentive structure as well: properties deemed to be riskier incur higher permit fees, but can move into a lower tier by cleaning up their inspection record or making proactive repairs.

However, the tier system has its limitations. First, tier assignments were based solely on how many inspections and violations a property had in the past, which leaves out a rich amount of information. Second, the city evaluates tier assignments for properties infrequently (every 3 to 6 years), and these adjustments require a great deal of expertise and manual work. We developed a predictive model to provide a more nuanced view of properties' violation risk over time, allowing for more efficient scheduling of inspections. https://dssg.uchicago.edu/2017/07/14/data-driven-inspections-for-safer-housing-in-san-jose-california/

In collaboration with **South Bend's Innovation and Public Works** departments, the Data Science for Good Fellowship used the city's data to develop a deeper understanding of the human behavior that characterizes water shut offs and quantify the scale of the problem and its consequences. This work informed the design of a behaviorally-driven intervention aimed at lowering the frequency of late payments and shut offs, which will be tested by the city this summer. https://dssg.uchicago.edu/2017/06/28/combining-data-and-behavioral-science-to-reduce-water-shut-offs/

A valid concern with using Crowd Science is privacy. You will need to ensure all of the data you provide to outside companies and data scientists does contain personally identifiable information (known as PII). Concerns around privacy are often used as a reason not to glean insights from data. The fear of letting PII data get out into the public outweighs the benefits for many NGOs.

Overcoming this fear will require focus regarding anonymizing your data. There are a number of techniques to utilize, starting with removing the PII information completely. Names; social security numbers; credit card information; phone numbers; and addresses are examples of PII.

When it comes to PII, you will discover there is a debate about the difference between sensitive-PII and non-sensitive-PII. Phone numbers and addresses are an example because they are listed in the phone book. But what about unlisted phone numbers and addresses?

The first question to ask is whether any of the PII data is needed for the project. If not, then don't use it. If location is important, consider using micro-aggregation, where addresses are aggregated at the zip code plus 4 level, which translates to an average of 4-7 households. You can then aggregate all other relevant data at this level.

Another technique is pseudonymization, a procedure for replacing PII fields with a pseudonym. For example you could use the field labels (First Name and Last Name) in place of the name in the field or you could replace the names with random ones. An important consideration is whether omitting the data will cause the results to be degraded to the point they are no longer useful.

Keep in mind what the data scientists need is not necessarily what you will need to make the information actionable. For instance, you can provide an ID instead of a person's name and then connect the data science results back to your database using the ID without the PII data ever leaving your organization.

You might be surprised how little data it takes to figure out who a person is. One study concluded 87% of the U.S. population could be identified with just their gender, birthdate, and zip code. This is why all of this work needs to be overseen by your Data Governance team. They will decide what data can be sent for analysis and will evaluate the risk/reward for the organization. One of the biggest ROIs you will receive from a strong Data Governance program is being able to confidently take advantage of both internal and external expertise.

Crowd Science clearly can work, but what if you want to have your own data science team? Is it possible even on a tight budget? Yes, and here's how to make it happen.

The first place to start is your mission. It is the reason someone would turn down a better financial offer. Just like there are social entrepreneurs like me, there are social data scientists. An example is Susan Scrupski, Founder of Big Mountain Data (www.bigmountaindata.com), who has focused her firm on using data science to combat intimate partner violence.

When you place an ad for a data science position, don't make the mistake of making it sound just like the one Walmart or Netflix® would write. Focus on your mission's impact and on the types of projects your organization is working on. Emphasize the difference a person can make by joining your team. Appeal to their passions and their heart — not to their self-interests and wallet.

Another place to find your in-house cognitive talent is colleges and universities in your area. Through my work as an Advisory Board Member of the University of Central Florida Master of Science in Data Analytics (http://www.ce.ucf.edu/credit/master-data-analytics/) I have seen first-hand the amount of talent in our higher education system. And the great news for NGOs is they are ravenous for real data and interesting and impactful projects.

Reach out to your local institutions and see if you can partner to help your organization solve complex problems, and provide the institution with a source for real-world experience for their students. You may be able to get the work done for free or at intern rates, which are far below market value.

I am also the Co-Founder of a data science Meetup group in Tallahassee. Within a couple of weeks of starting the group we had nearly 200 members. For a city the size of Tallahassee that is an impressive number. If you are in a larger metro area you can expect to see multiple groups focused on different aspects of data science.

You and members of your team need to join these groups both as a learning experience and as potential recruiting opportunities. You also may discover the group would like to take on your data challenges as a project for nothing more than beer and pizza.

If budget was one of your reasons for not embracing Cognitive Computing, you need to find another one. Just like every other aspect of an NGO, it will take some imagination and creativity to make it happen, but happen it will.

09

The Cognitive Organization

"If you want to change the culture, you will have to start by
changing the organization."

—Mary Douglas

What will the Cognitive Organization look like? Just as we had
to make room on the org chart for the emergence of
technology, we will need to re-imagine our organization to
make a place for cognitive applications. This will not be easy.
It was not too long ago the idea of a Chief Technology Officer
was not even a thought in leadership's mind.

We are already seeing the political battle lines being drawn as
Cognitive Computing begins to change organizations. Is

Cognitive Computing just another technology, like an Oracle® database, or is it something more?

This debate is about much more than just technology. It is also about who will control the data and the insights it provides.

Too often tech infrastructure is confused with what it empowers. The power company doesn't care about the microwave and vice versa. At the same time data is seen as a cost and a commodity to be managed by others when in fact it is an invaluable business asset needing direct senior management.

Cognitive Computing has three essential elements: 1) Technology 2) Data and 3) and Insights. No part can function without the other two. Someone needs to own each area, and someone needs to manage the coordination of the three.

Here are the roles and responsibilities of each position:

Chief Cognitive Officer

Senior Manager of Cognitive Computing responsible for operations with the goal of delivering actionable insights to the organization.

Chief Data Officer

Responsibilities include internal and external data acquisition; governance; data curation; quality assurance; and transformation of data into information.

Chief Technology Officer

Responsibilities include infrastructure; security; reliability; and delivery of data, information, and insights.

Chief Insight Officer

Responsibilities include data science; machine learning; and deep learning to transform data into actionable insights.

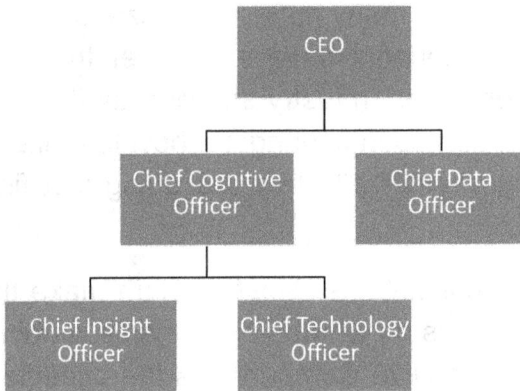

Note the Chief Data Officer reports directly to the CEO. This is the only way to ensure your Data Governance policies are not unduly influenced by the insight needs of the organization or the desires of the technical team. Risks associated with data are at the organizational level, so senior management has to have someone who is focused on mitigating those risks while also supporting the operations and mission of the organization.

This model ensures Cognitive Computing is not subsumed into the tech department, which will hinder its ability to be integrated into the organization. The Chief Insight Officer and the Chief Data Officer will act as bridges for subject matter experts to train and interact with the technology. They will need to have interpersonal skills to go with their data science and data management abilities.

This is a moment for leadership to establish an insight-driven culture where everyone feels they have input. Cognitive Computing is about inclusion, not exclusion. Establishing an organizational structure to support this culture shift will soon be critical to your success.

When we are in times of transition it can be challenging, as we are not able to find many other organizations doing it. An interesting phenomenon I have seen over the years is it is rarely the leaders in an industry showing us the way forward. They are too caught up in figuring out how to make the current reality last longer. After all, they are doing just fine the way things are.

Also, beware of consultants. Most of them make their money helping organizations do what everyone else is doing. This will be great when Cognitive Computing is the accepted practice, but until then they will say "it's too new" or "it's not for your organization," and then have you do it the way it has always been done AKA "best practices."

Look for innovation to come from unexpected corners. It could be a new organization or a small to mid-size one where the risk of doing something new is not being judged against current success.

Find places where Cognitive Computing is solving a critical problem. Remember the first users of FedEx were the medical and legal communities, not all of us procrastinators. It's not surprising Watson's first post-*Jeopardy!* application was in the cancer field. There is no more pressing issue than death.

Beyond healthcare, we are seeing education adopt cognitive solutions to discover the most successful journeys for students. This, in turn, has led to the idea of a digital campus focused on providing the best experience for students from admissions to being an alum.

Disaster relief organizations need a multitude of insights from weather to demographics to supply delivery. Environmental groups are confronted with ever-changing climate, political, and economic realities. The arts, just like Netflix, must have its finger firmly on the pulse of the communities it serves. Social service groups are asked to take on more and more responsibility, and those responsibilities are impacted by a variety of factors both in and out of its control.

As we envision the Cognitive Organization, keep in mind this is not just about bringing technology and techniques into your operations. Truly cognitive organizations will be about harnessing the cognitive computers on our shoulders as well as the ones in our phones; computers; LCDs; and the AI assistant asking about your day.

Cognitive Balanced Scorecard

When evaluating how your organization is performing there are two conflicting goals: Effective mission and efficient operation. Using the same measures for both does not work because the underlying forces are so different.

Missions are rarely a linear process. Transforming a high school graduate into a college educated individual capable of succeeding in today's complex world requires a combination of academic; social; and financial factors. Success or failure could come down to something as simple as whether they are living on campus or off. The complexity is even greater with social service organizations where the human factors can range from mental health to substance abuse.

Operations acts a bit more like a for-profit business, but even here there are unique elements such as pricing (gift levels) being variable for the same product (your mission). No other industry can charge $1 or $1,000,000 for essentially the same thing based on the capacity of the person to pay.

Another challenge is your costs are not necessarily going to happen in the same reporting period as your revenue. Major gifts can take 18-36 months. This means in some cases expenses for only one year out of three are taken into account during the year the gift is received. This is just one of the reasons cost-of-fundraising is a poor measurement of performance.

For NGOs, customers are donors and mission beneficiaries. Internal business processes are related to both fundraising and the mission. For this reason your organization needs a balanced scorecard for both aspects.

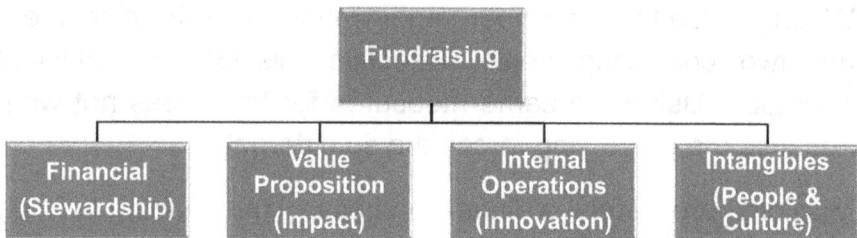

```
                          ┌─────────────────┐
                          │     Mission     │
                          └─────────────────┘
        ┌───────────┬───────────┼───────────┬───────────┐
┌───────────────┐ ┌───────────────┐ ┌───────────────┐ ┌───────────────┐
│   Financial   │ │     Value     │ │   Internal    │ │  Intangibles  │
│  (Efficiency) │ │  Proposition  │ │  Operations   │ │   (People &   │
│               │ │(Effectiveness)│ │  (Innovation) │ │    Culture)   │
└───────────────┘ └───────────────┘ └───────────────┘ └───────────────┘
```

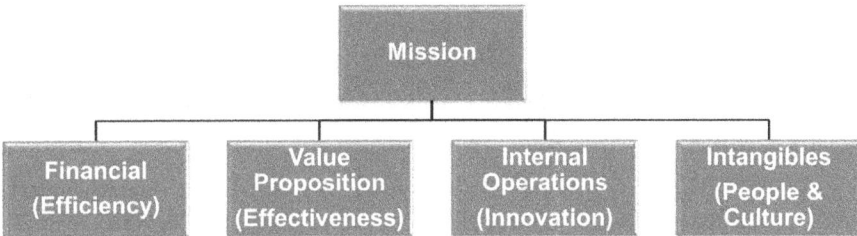

Where cognitive comes into the mix is to provide the qualitative data to understand the quantitative results. How else are you going to measure *satisfaction* of your donors with your use of their donations? Where else in your strategic initiatives are there outcomes driven by qualitative forces? Employee retention may be one, and that leads right back to satisfaction.

Are you measuring the *happiness* of your mission beneficiaries? They may have a degree, but are unhappy with the career to which it led them. They may love the design of your new performing arts center, but do not like the music selection. Perhaps you are meeting your fundraising goals yet losing donors because your focus on major gifts has been at the expense of your mid-level program.

Let's look at some of the ways you can use your Insight Reservoir:

Review Sites – Glassdoor (www.glassdoor.com) is an example of sites providing anonymous reviews of companies and organizations from employees and former employees. While you might think it's just a place to complain, what you will discover is a mix of insights into what your organization is doing well and what it is not.

Social Media – Learn what your constituency thinks about your mission, and about your fundraising. You may have a strategic goal of rebranding your organization to be perceived a certain way. Social media is an excellent source of information on whether you are succeeding or not.

Economic Data – Your scorecard is going to be influenced by forces beyond your control. Depending on the scope of your organization, these may be local or global. Bringing relevant economic data into the mix will ensure your results are measured in the context of the realities you are facing.

Survey Data – You can develop open-ended questions for your constituents designed to inform your scorecard. Rather than long "on a scale of 1 – 10" questionnaires, create short 2-3 open-ended question surveys which are sent to key constituent groups. You can use Cluster Analytics to develop your groups.

As you incorporate these insights into your scorecard be sure to include trends over time. Where you are is important, but understanding where you are going is even more critical. You may have low donor retention yet if it has risen dramatically over the last few years as you implemented new initiatives to increase donor retention — that is the story, not the current number.

Of course, there will be times the news is not good. It will be tempting to hide or at least obscure these insights. Fight the temptation. Instead, use Cognitive Computing to help explain why it is happening. Another benefit of honesty about your failures is it will lend credibility to your stories of success.

Keep in mind, people are conditioned to think data has been manipulated to prove the speaker's point; highlight success; and hide failure. Transparency is one of the hallmarks of a Cognitive Organization. Without it, the value of insights will be greatly diminished as people seek to control, rather than to harness, their value.

Linking your Insight Reservoir to your balanced scorecard will not only enrich your results, it will help your stakeholders understand the value of the investment. It may also lead to more engagement with the reservoir as well as additional investment.

Agent C – Your 24/7/365 Cognitive Assistant

Until now technology has always had an on/off switch. There was a certain comfort in knowing while the machine might be capable of doing something you could not, it needed you to give it power. A fully implemented cognitive platform takes that power away because it is designed to never stop. 24/7/365 operation is fundamental to its operation. It may have peaks and valleys of activity, but just like us it never turns off (of course, when we do, it's forever).

My first experience with this always-on concept was in the late 90s as the World Wide Web was exponentially expanding the amount of information available via the Internet. I was looking for a way to search across the web using one search rather than having to go from site to site. This is called metasearch.

I found a software package designed to do exactly what I wanted, take a search from site to site and return the information. Further, it would de-dupe the information because sites often had the same data. The user experience was novel at the time – you gave your search criteria to a hound dog, and then told it to fetch your information. If you wanted you could watch it go from place to place or you could just do something else. It didn't matter if you turned off your computer; it kept searching the web.

When it returned you would review the information found, and, in essence, say "good girl or bad girl" and, from this, the application learned what kind of data (or site) I liked and what I did not. Then Google came along, and this was a distant memory as we now had the most powerful metasearch engine for free.

Between then and now we have been conditioned to ask our phone for answers, and can even have AI schedule meetings for us. While adoption is far from universal, the idea of a Cognitive Assistant is taking shape.

What I see happening is the merging of what we now know as search and queries with the ability to understand both the person asking and domain it is searching within. These will be Cognitive Agents (hopefully yours will be more 007 than Maxwell Smart).

Think of the questions you would like answers to at any given time. What is happening with our fundraising campaign? How is our new project going? We have been conditioned to only ask questions periodically because it requires human capital to get the answers.

Would you like to have answers to your questions whenever you need (or even want) to know them? Of course, you do.

In the cognitive age this is not only possible, it is central to the whole concept. Would you want your brain to not answer when you ask yourself what you want for dinner? Imagine being told your brain is off-line when asking it to recall the directions to your house.

If your cognitive computer is always on, then it can always be asking your questions. Your Cognitive Agent just needs to know what you want to know, and also when you want to know.

For example, you might ask, "Tell me if there are any personnel changes that might impact my capital campaign." This might return the fact a leadership giving fundraiser resigned, placing his portfolio at risk. On the mission side, it might be a high-performing employment counselor whose exit may cause a drop in employment for your low-income training program.

Notice I didn't ask "How much have we raised?" or "What are our employment numbers?" We do a reasonably good job of capturing that data in KPI dashboards. Cognitive can easily pick those up, but what you really want are the drivers behind the numbers.

Sometimes what you want to know requires a question you don't want to ask your human colleagues. This is where the Cognitive Agent can provide a way to discover insights without alarming anyone.

To picture how all this will work, imagine the Insight Reservoir and these Cognitive Agents as little submersibles able to go anywhere. Inside are your questions, and the background knowledge needed to know what to look for. Initially, your answers may be good but not great. This is where you will say "good agent" and "bad agent."

Over time your Cognitive Agents will become intelligent enough to suggest questions, and also to draw inferences to share with you. At some point not too long from now we will wonder how we ever lived without our little CAs. You can fall asleep knowing while you dream about the big meeting tomorrow, your CA is gathering the latest insights and will share them with you as soon as you look at your phone in the morning.

The Leaders View – 360 Isn't Good Enough

The "360-degree view" has been the Holy Grail for organizational leadership. Vendors and consultants promised a complete picture of your relationships regardless of whether your interactions were on-line or off-line. Millions of dollars, and countless hours, have been spent making the 360-degree view a reality.

For fundraisers, this meant bringing together all of the elements of a successful campaign. For other departmental silos, it delivered the view they needed to get their job done. Leaders began to receive more complete data from their direct reports, and for a while it seemed as if 360 had helped organizations do a 180 in terms of having the data needed to make better decisions.

But there was a problem – in fact, many problems.

First, the 360-degree view was essentially a new way to look at all the structured data found in a traditional database of record (known as a donor management system for fundraisers). With 80%+ of today's data being unstructured, this meant the view was far from complete.

Second, it was a 360-degree view of *each silo – not of the organization*.

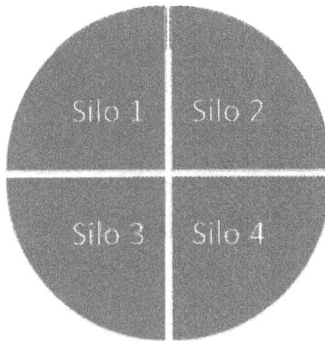

For leadership, this means they are receiving varying degrees of information from each silo, at various times and in various formats. It is like a ship's captain receiving reports from each deck at different times – some more complete than others. The captain makes the decision to continue the journey not realizing the refrigerator is broken and the food will spoil soon because that department is late with their report or perhaps doesn't want the captain to know there is a problem.

For the President of a college this might be the cancellation of season football tickets by an alum she is counting on to support the capital campaign.

The head of a homeless shelter could be asked to be part of a tour of the shelter for a large donor who he discovers, while on the tour, volunteers there once a week, which would have been best known prior to the tour.

A headmaster of a private school calls on a prospective donor, not knowing their daughter just received a very negative report from one of her teachers.

These are just the problems associated with structured data. What about all of the unstructured data? That wasn't even considered in the 360-degree view. Are leaders able to see not only transactional elements of impact reports, but also analysis of written responses?

With Cognitive Computing, leadership has a multi-dimensional view across their organization encompassing all of their structured and unstructured data. You have the *whys* behind the *whats*. This is what leaders need to be successful in a hyper-dynamic and hyper-competitive marketplace.

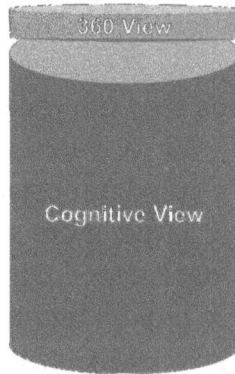

Leaders also enjoy unfettered access to data, which puts them in a position to see cross-organizational insights without the filters middle-managers often apply. The President of a hospital can see both medical and donor reports. A college president can see admissions, student, and alumni/ae data. The head of an NGO can see volunteer, member, donor, and beneficiary information.

Because of regulations, department heads may only have abbreviated or anonymized data, which is better than nothing, but can leave blind spots in their insight.

Leaders understand what they don't know can, and does, hurt them. For them viewing only what their silo managers can see is not enough in the dynamic world in which they operate. They need a cross-organizational view to confidently, and safely, set a course for sustainability.

If you are a leader, then hopefully you are anxious to get this cognitive view. If you are not the leader, then you may be becoming anxious thinking of leadership getting a real-time look at your area.

As much as we complain about technology not working, there has been a certain comfort in knowing it was not easy to obtain information. It has come in handy to be able to poke holes in any analysis by saying it was not complete or didn't consider a qualitative measure.

This has slowed how data, and especially insights, travel within an organization. It has invited people to manage, manipulate, and, at times, make the insights go away before leadership has a chance to see them.

The phrase "nothing lies like statistics" really should be "it's easy to lie given access to the raw data and sufficient time to shape it to support whatever one desires to be reality."

An example of this is the decades' long battle between direct marketers and major gift fundraisers. I know well, and am related to, some of the leaders in both areas. To the direct marketer, the best way to raise a million dollars is to get 100,000 people to give you $10. This then gives you a 100,000 donors to upgrade, and in an organization where size matters for advocacy, those 100,000 people give your organization power beyond money.

A major gift fundraiser would say why not just raise a million dollars from one person? And they would add, please don't send them annoying letters asking for $10.

The tension between these two camps has led to some interesting moments in my career. The donors I screened were often acquired by direct mail, yet the wealth information I found was destined for major gifts. In one organization, controlled by the direct mail folks, they actually said their major gifts department could not ask for any money from the millionaires and billionaires I had discovered. They could only do what amounted to a stewardship letter. Funny thing is, some of those donors still made major gifts.

Though I have spent most of my time with major gifts, I want to be clear — I absolutely believe in direct marketing of all kinds. Without it there would be no constituency from which to find wealthy donors. I also don't believe you should take people out of the direct marketing program once they are discovered. They should receive a different type of communication, but you don't stop asking for the annual gift because there is hope of a big gift down the line. I have seen too many instances of "assigned, but no activity" where the prospect was so special everyone was afraid to do anything.

One final note on this subject, beware of expert bias. I have seen articles, presentations, and blog posts asserting identifying wealth in your database doesn't work because it doesn't move the needle. When you look deeper, you realize they studied organizations with an immature, or no, major gift program. The best data in the world is worthless without a new action based on the insights it brings.

Conversely, you have people who profess direct marketing is not effective, and they use as their proof organizations with poor direct marketing programs and a strong major gifts program. Next time you read or hear an expert's advice, look at their bio. You will quickly realize the bias they are bringing to their opinions.

A leader can break through all this turf protection by seeing a donor's journey from beginning to end (which can happen with planned giving). The Insight Reservoir flattens the organization. There are no silos; there are no agendas; there are no politics; there are no implied biases; and leadership can access these insights anytime they want.

A central challenge of Cognitive Computing will be how to make insights available to leadership in ways they can not only understand, but also via deliverables or tools which don't overwhelm them. This will be easier for leaders who are comfortable with current technology, but all leaders need to be involved in the questions and need to have an understanding of how the insights are being created. However, they do not need to have a deep understanding of how it is being accomplished or all of the gritty details about the data.

Three areas leadership can bring their considerable power to bear on are breaking through political barriers regarding data ownership; finding resources to support Cognitive Computing; and making acting on the insights derived high priorities.

While some leaders are going to adopt cognitive on their own, others will be encouraged (or forced) by their boards. The men and women who sit on these boards are often from the business world. They are seeing firsthand how transformative this technology can be for their companies. As they gain a deeper view in their professional world experiences, they will come to expect it from their philanthropic activities.

How can you determine if your organization is ready? One indication is you currently are receiving quality views of your structured data. You may have a data warehouse and/or a reporting database that gives you the data you need. You most likely have at least one data scientist who is comfortable with traditional analytics. These are all part of a strong foundation for an Insight Reservoir.

In this situation, you will be able to integrate your current data warehouse into the Insight Reservoir, and your data scientist will be able to acquire the Big Data science skills relatively easily.

If, on the other hand, you are not happy with your current reporting system, then there will be more work because you will need to solve both the structured and unstructured data problems. You also most likely do not have the skills in-house.

In this case, your first step should be to bring in a consultant to review your organization and help you create a plan. Even when you think you have everything together, this can still be a good idea because there are going to be many important decisions, such as what you should do on-premise and what you should do in the cloud. You also need to look at staffing.

Another key factor is an honest assessment of your organization's ability to act on insights. This starts with people and budget. People are not just about numbers. It is also about their abilities. Cognitive Computing can power a successful omni-channel marketing program, but not without people who know how to implement one. Is there someone, or if you are lucky multiple people, with the data science skills to work with an Insight Reservoir?

The answer may be a combination of in-house and outside expertise. Cognitive Computing is available in a Platform-as-a-Service model, so be smart about how much you build vs rent.

A real danger is to create too much home-brewed technology, which while, it may work, leaves you with no support and does not keep up with the latest capabilities. This is what happened with relational databases in the 80s and 90s. Beware the engineer who tells you, "I can do that!" What they don't tell you is how long it will take, and that once they are gone you will not be able to support it.

As technology found its way into organizations starting in the 80s, leaders could put distance between themselves and all the gadgets by deferring to IT. In the Cognitive Computing age this not an option. This wave of technology is not just about getting things done more efficiently; it is about integrating the tech into the organization's operational fabric. If a leader fails to be actively involved in this process, they run the same risk as if they were not involved in the hiring of key personnel.

The 360-degree view is being replaced with a 3-D view. For leaders it may feel a bit disconcerting at first, but in time you will marvel at how anyone guided an organization without it.

Our Cognitive Colleagues

©2017 Max Lawson

Can we become friends, or even acquaintances, with non-humans? Based on all the cat videos and dog photo shoots coming across my Facebook feed as well as the many dogs and cats I have had in my life, the answer is an emphatic yes. In the case of dogs, we have trained them to retrieve balls, help us across the street, hunt, find explosives, and alert diabetics when blood sugar levels fall (https://can-do-canines.org/our-dogs/ourdogs/diabetes-assist-dogs/). With cats, we have just taught them to tolerate us.

We are clearly comfortable giving animals consequential tasks. Even more importantly, we are willing to form a relationship with the animal. The K9 officer loves his or her dog the same way we do when not on duty, as does the soldier working with his dog to find IEDs. This bond is critical to the dog being able to perform its tasks.

Can we form this same type of relationship with a cognitive application? Will it need to be lovable like a pet? Will it need to know we are the Master?

We are moving beyond technology simply telling us the best route for our trip, the answers to basic questions, and ordering a pizza. Siri, Alexa, and countless map apps have made these functions part of our new normal.

What has not been normalized is the idea of technology being seen as a colleague, something with which, or with whom, we having a working relationship. While this concept has been in sci-fi movies for decades, it was always in the not-quite-accessible future.

There has always been a safe distance between humans and technology. Until the personal computer, it was controlled by governments, large corporations, and a relatively small number of people with the expertise to use it. For the vast majority of us, technology was unapproachable, something to be amazed by, but not understood.

Over the last 40 years we have become comfortable harnessing technology to perform basic tasks, provide us business intelligence, and in the last decade able to trust it enough to make even critical business decisions using prescriptive analytics.

Cognitive Computing is going to require us to become far more intimate with our technology. We will need to move from *personal technology* to *relationship technology*. Just as an employee needs to understand not only how to do their job, but also form relationships with co-workers, Cognitive Computing will need to be allowed into your inner circle to succeed.

To make this a reality companies are trying to *humanize* technology by showing people having a conversation with it. Humor is another critical element. The ads with Watson talking about its relationship with a coffee maker are meant not just to make you smile, but to also lower your resistance to having Watson become an accepted member of your team.

One area Cognitive Computing needs to work on is voice quality. It is time to end the classic computer speak which is anything but natural. We are not going to interact with something sounding so strange on a day-to-day basis, especially when it is trying to have a conversation with us.

Another obstacle is physical appearance. Will we accept a robot sitting next to us in the conference room? Will we interact with it as if it was human? Robotic movements are becoming more fluid, and they certainly look less and less like the robots of old. Add a latex mask, and you could get pretty close to something human like.

For Cognitive Computing to be integrated into your organization, I don't believe we need to wait until we have human-like androids. As long as we can humanize the voice, we can assimilate Cognitive Assistants into our work life.

What gives me confidence is the proliferation of virtual meetings, video conferencing, and group work platforms. When these were first introduced they were met with heavy resistance. People believed you had to be face-to-face to get anything done.

The idea of everyone not being in the same office was heresy. Now we know working from home can actually be more productive than being in a physical office building, and virtual teams are the norm. In this environment having a cognitive member of your virtual team is not such a leap.

©2017 Max Lawson

Another way to understand how this can work is how our organization utilizes consultants. I'm not saying consultants are robots, but they are brought in for their expertise, assimilated into the organization's operations, and allowed to have a voice at the table. This is despite the consultant often not spending a great deal of time face-to-face with the people he or she is advising.

The Cognitive Consultant (CC) may provide a path for Cognitive Computing adoption. We can bring the CC into a meeting much as we would a consultant, to help us work on especially difficult situations. Over time we will become comfortable, and the CC will morph into a Cognitive Colleague.

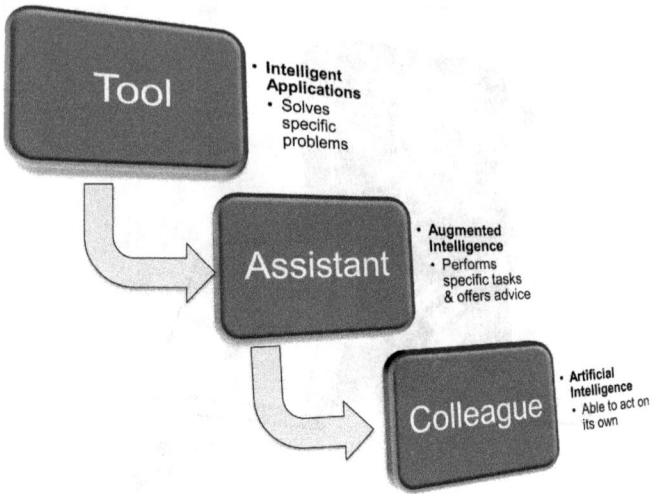

- **Tool**
  - Intelligent Applications
  - Solves specific problems

- **Assistant**
  - Augmented Intelligence
  - Performs specific tasks & offers advice

- **Colleague**
  - Artificial Intelligence
  - Able to act on its own

10

Being Wrong on the Way to Being Right

"I have not failed. I've just found 10,000 ways that won't work."

—Thomas Edison

"Wrong answer!" No one likes to hear that. From school to the conference room, we dread the words.

This conditioning has made us try very hard to never answer a question, or make a statement, unless you have a high confidence you are right. That is one reason you don't hear a lot of new ideas in meetings. No one wants to be told they are wrong.

Computers don't care if they are wrong. In fact, sometimes I think they enjoy it, or at least find amusement giving us the wrong answer. In the cognitive age, we need to harness this child-like quality to ask all the questions to which we don't have answers or we suspect our answers are not as good as we think they are.

One reason there are so few Edisons is there are few of us who are willing to be perceived as failing. Yes, Edison was a genius, but without his unwavering belief that every wrong experiment was just a stepping-stone to success, the world would be a lot darker.

Embracing Edison's philosophy is a key step in your cognitive journey. You must accept the iterative process of this new technology. On day one it is not going to be all-knowing and all-seeing.

If you have not seen the movie *War Games*, take a trip back to the 80s and see how this technology is applied to nuclear warfare. I don't want to be a spoiler, but what it will show is two things: 1) Cognitive systems need time to learn; and 2) Don't connect it to the missiles until the learning is complete (or at least complete enough to feel safe).

In your world, think of missiles as important aspects of your mission and operations.

Don't directly connect insights to actions until you are comfortable the decisions are as good as or better than what you can do on your own. Be careful here because you could create a bar the technology never gets over by demanding it never be wrong.

You may remember the story of the man who discovered Target had predicted his daughter was having a baby, the sex of the baby, and the month she was giving birth. This freaked everyone out as they imagined how this could be done. The facts are this was based on very few purchases including a blue round rug and stretch-mark cream. Using this in combination with the dates of purchase, and historical data on women who had then purchased post-baby items, they had an 87% probability of being right.

That means there was a 13% chance they were wrong. It could have been a sister, aunt, mother, father, or friend.

Now think about decisions you make every day. Are you 100% sure of every decision? Have you ever been wrong?

Once you embrace the power of being wrong, you constantly challenge your Cognitive Assistant with questions. You don't look at wrong answers as signs it is not working, but rather as signs it is working towards the answer.

We now all can be Edison. Our laboratory is the Insight Reservoir. Imagine all the light in the world as philanthropy frees itself from having to always be right.

Accidental Insights

The history of invention is filled with stories of accidental discoveries, from failing to create plastic gun sights in WWII leading to Super Glue® to Post-it® notes which began as a failed attempt to create large sheets of paper with adhesive on the back.

As your Insight Reservoir grows, be on the lookout for your accidental discoveries. They will be hiding in the answers to seemingly unrelated questions.

An example might be a question about your direct mail program – Why are fewer people making donations by mail? You may discover correlations between changes in the demographics of your constituents or the type and frequency of the mail you are sending. You may also discover success of your online giving initiative during the same period as the decline in donations received from your direct mail campaign.

If you stayed focused on your question – Why are fewer people making donations by mail? — you may miss a connection between your mail and online giving. Are the people receiving the mail deciding to make their gift online rather than putting a check in the envelope and mailing it? You could determine this by looking at the timing of mail campaigns and the timing of online gifts.

In the academic world, you might be looking at ways to increase the number of out-of-state students. At the same time, you have been studying housing. The study reveals a high level of dissatisfaction with dorms. You now look at comments from your recruiters about what out-of-state students are saying to them when they come to campus, and you see this is a core reason you are not being successful. No amount of marketing is going to help until you solve the housing issue.

Armed with these insights it's time for a housing campaign, and you will be able to share the insights with prospective donors to help them understand why this is a priority. Here in Tallahassee, FSU College Town, complete with upscale student apartments, restaurants, and entertainment, was developed to attract South Florida and out-of-state students.

The secret to being purposeful about discovering accidental insights is creating an environment where you are organization-focused. This means data science projects are designed to not only provide insights to the people asking the questions at hand, but also to contribute to the overall organizational insight.

The Insight Reservoir will facilitate the process by ensuring all of your insights are centralized. This means when discoveries are made in one part of your organization, other parts can immediately apply them to their questions.

This is another example of why machine learning is so powerful. Unlike a query, report, or static-score, Cognitive Computing is constantly challenging its own assumptions. And the best part is you will not have to hope the people in another department remember to tell you about the helpful (for your area) insight upon which they accidentally stumbled.

Deductive vs. Inductive Reasoning

> "All truths are easy to understand once they are discovered; the point is to discover them."
>
> —Galileo

We can all remember our first science project where we had to form a hypothesis and then do an experiment that either proved or disproved it. This is called deductive reasoning. This is how we approach classic analytics. We state, "Baby boomers are not giving to their alma maters because they are focused on organizations involved in more pressing areas such as poverty and the environment." In essence we think we know the answer before we ask it.

With inductive reasoning we start with a very broad question, which acknowledges we don't have a hypothesis. We simply ask, "Why are Baby Boomers not giving to our institution?" We can then open that up by asking, "Why are people not giving to our organization?"

This open-ended, inductive approach is essential to gleaning new insights from Big Data. You may discover, as weather experts did with the severity of Atlantic hurricane seasons, that it has nothing to do with what you thought it did. By asking "Why are some years more severe than others" they allowed the water temperature in the Pacific to be a data point, and that led to the El Niño and La Niña effect. With deductive reasoning they would have asked, "What factors in the Atlantic are impacting the severity of the hurricane seasons?" It would have been reasonable since every other aspect of an Atlantic hurricane has something to do with data from the Atlantic region.

This is also where you find the basis for "nothing lies like statistics." With deductive reasoning you are invested in your hypothesis, and you can find a way to make the data prove your point. With inductive reasoning you don't create a bias towards one conclusion over another. You can only have the thrill of discovering the earth revolves around the sun or the earth is round if you don't assume the answer is what you expected.

At your next staff meeting challenge the group to come up with questions to which your organization needs answers. Separate them into deductive and inductive questions. Then take the deductive questions and work on turning them into inductive ones.

Deductive	Inductive
Are women not supporting our mission because we don't have programs that interest them?	Why do people support us?

The deductive question is typical when people in the organization want the outcome to align with their own observations and beliefs. The inductive question leaves open the possibility the lack of support from women has nothing to do with gender. Perhaps the problem is the female prospect pool is smaller, or has less potential, than the male pool or is it a matter of always counting a couple's gift as having been received from the male of the household. You may find the results you are seeing align very well with what is possible. Also note the inductive question does not only look at the people not supporting the organization. Only by looking at both people who do, and people who don't, support your organization will you discover root causes of your performance.

Becoming inductive frees you to go where the insights take you. Of course, you may encounter a lot of resistance because your insights upset the fundamental beliefs of your organization. Hopefully you will not suffer the fate of Galileo, and spend your last days under house arrest.

11

Personalized Fundraising for All

> "Philanthropy is the mystical mingling of a joyous
> giver, an artful asker, and a grateful recipient."
>
> —Douglas M. Lawson

My introduction to fundraising came at an early age. My father, Douglas Lawson, had moved from Chaplain to Dean of Men to Vice President for Development at Randolph-Macon College. He was also an ordained Methodist minister. You could say I was exposed to fundraising from the beginning of my life as the collection plate passed by my mom as she held me while my dad gave his sermon, but I truly became aware of professional fundraising after my parents divorced.

Dad moved to D.C. to first be part of the Urban Coalition with John Gardner. He was invited to be part of that extraordinary team of fundraisers because of his work helping to integrate Randolph-Macon in the late 60s. Haywood "Hap" A. Payne, Jr., class of '68, was the pioneer, and he went on to be a top executive with Chevron.

After the Urban Coalition he founded a fundraising firm with Walker Williams, and Lawson & Williams would go on to raise millions of dollars for causes across the country and around the world. It was in the early 70s I was given a summer job "editing" Foundation Research Service, a multi-binder set of profiles on the top 1,000 private foundations.

My job was to find the pages the printer had printed upside down then turn them; three-hold punch them; and put them back. Unbelievably that experience made me hungry for more, and the next year I was the assistant editor on a new book – The Foundation 500. In six weeks my brother, Sam (the first editor), and I put out a book that not only categorized every grant, but also identified in which States the Foundations gave.

The technological revolution for The Foundation 500 was the wide-carriage IBM Selectric, which allowed us to create the final copy for the printer (all pages right-side up). Without this amazing typewriter we would have had to typeset the book, a far more costly option.

During all of this I would listen to my dad and Walker discussing their latest fundraising campaigns. Their approach was very personal, and even involved them making the ask for organizations without a good asker. Today, saying "People give to people, not organizations" is met with "of course," but back then one-to-one fundraising was revolutionary. This personal approach led my Dad to make friends with many of the people from whom he raised money and that led to me meeting these philanthropists.

Being a teenager when you meet people of wealth is a good thing. I couldn't care less about their money. They were just interesting (or sometimes not) people. It also taught me the value of giving. The happiest people with money I have ever met are the most generous.

My dad very much wanted me to join him as a consultant, and I even started down that path before I succumbed to my true love – data and technology.

Later when my first company, The Information Prospector, was starting to make some noise in the industry I learned another valuable lesson about fundraising.

I was at a conference where I was exhibiting and speaking. A lot of our clients were there, and a number of us sat at a table together for lunch. This particular conference hotel was connected to a hospital, and the dining room was shared.

As I was enjoying the feeling an entrepreneur gets when his or her idea is gaining traction in the market, I saw a young girl walk in. She was pulling her IV stand and had no hair. Suddenly the voices I had been listening to so attentively went silent, and it felt like only she and I were in the room. At that moment I realized why I was doing what I was doing. It was not about the business, money, or small amount of fame. It was to help put hair back on her head.

I tell you all of this because one way to not embrace a new technology is to say "they don't understand this is about people, not computers and impersonal communications."

I understand that fact deeply. My mom was also a fundraiser (she raised the funds for the documentary *City Out of Wilderness* (among many other causes), and my grandparents were lay leaders in the church. My wife (and editor) has been in the field for the last two decades.

The de-personalization of fundraising by the overuse of scores and technology is real. The poor retention of donors is a symptom of our inability to scale one-to-one fundraising and especially what happens after a gift.

At the same time, data and technology have transformed our ability to scale one-to-many outreach. From email to crowdfunding, we can reach millions of people with segmented messaging, and giving is as easy as pressing a donate button. Despite this success, it still can leave one with a feeling of hollowness compared to personal contact.

Cognitive Computing is the bridge between technology and the individual. Fully realized, it will enable organizations to see every constituent as a market of one. This will lead to life-long relationships benefiting both the organization and the supporter.

So don't despair the end of personal fundraising. The only difference will be our ability to scale it.

Virtual Engagement – Mixing Realities

Philanthropy is about faith, even when the organization being supported has nothing to do with religion.

I was reminded of this when an organization with which I worked sent videos (actual physical videotapes) to donors who had supported their relief efforts after the Indonesian tsunami. The responses from donors underscored how much faith plays a role in giving:

> Many donors said they were thrilled to see their giving in action, and many gave another gift even though a solicitation was not part of the package.

> Other donors commented they had not been "sure" their gifts actually were used to help people. Despite this uncertainty, they made the gift in the hope the organization would fulfill their mission. That is the definition of faith.

We can, and do, rely on the strong faith of our donors. There is nothing inherently wrong with this, especially if you are a faith-based organization. The challenge is without reinforcement, like the videos from the relief organization, faith can be eroded over time as the donor begins to doubt the impact of their giving.

Stewardship's role is to ensure donors not only keep the faith, but turn it into life-long relationship beneficial to the organization, the donor, and the mission's beneficiaries. A key element of stewardship is providing concrete examples of impact. Schools send stories of scholarship students whose lives have been changed by the gift of education. Social service organizations show how families in distress are given the help they need to be secure. Arts organizations demonstrate how their activities lead to a stronger community.

Today, delivering these stories has become much easier as you can provide rich digital content on your website and via social media outlets. Because of this ease, and lower cost, the amount of stewardship content has exploded. Much as we are overwhelmed with advertising for products, we are becoming overwhelmed with social good.

For organizations this means it is growing harder to establish a unique identity. It seems there are an infinite number of organizations working on saving animals and the planet. Every hospital has excellent health care, and every school provides an excellent education.

How can you break through to tell the world why your organization's work deserves support?

Enter virtual reality (VR). You most likely know of the leader in VR, Oculus Rift®, a company now owned by Facebook. They offer headsets to visually immerse users in a three dimensional world. From taking a ride on a roller coaster to visiting the Great Wall of China, this technology can take you wherever (and whenever) an experience creator imagines.

Are there aspects of your mission which lend itself to VR, such as a tour of your facilities? What about your work in the field? Could you deepen the understanding of your mission by giving donors, and prospective donors, a virtual journey through your impact?

Now think about what it could mean if donors were brought together with people delivering your mission and those benefiting from it. Facebook paid $2 billion for Oculus Rift because it saw the potential for delivering virtual experiences that could be shared by its users.

VR is not a 3-D movie where you sit watching passively. You control your experience as you move through the virtual world. You also are able to join others who are in the virtual world with you. As I write this, Facebook has announced Facebook Spaces. FB users will be able to create events for themselves and their friends. They will also be able to attend concerts together.

Colleges are already giving virtual tours on their websites - https://college.harvard.edu/admissions/visit/virtual-tour. What you are seeing today is just the beginning of virtual experiences. Performing and fine arts organizations have access to a global audience, and healthcare can take volunteers and prospective donors on virtual tours of their facilities (and facilities they would like us to help create with our donations).

As VR becomes more commonplace, I can see fundraisers using virtual experiences as part of their cultivation as well as NGOs using VR for stewardship. Donors will no longer have to look at pictures or a campaign video. They can *see*, and even think they are *touching*, the new homeless shelter. Donors can ask questions of experts as they visit a village your organization is helping to provide clean drinking water.

You may see all of this as impossible because you lack the money or staff to produce and support VR. Remember, Facebook (and there will be many more) are going to provide platforms for creating and delivering VR. You are also going to be able to find volunteers with the skills to produce the content (reach out to colleges in your area) who would love to have your organization as a reference for their careers.

For larger organizations it may well be worth investing in having VR talent on staff, just as you now employ social media experts.

Just as VR comes of age, augmented reality (AR), has entered the market and may well play an even bigger role. The way you may know AR is Pokemon® Go!. Whether you have played it or not, you have most likely seen someone playing the game as they walk through a park or down a street.

What Pokemon Go! did was layer a game on top of the real-world by using the camera on your cell phone. You are alerted to nearby creatures, which you capture using Poke Balls found at Poke Stops. Those Poke Stops are often stores, restaurants, and public parks.

Facebook Spaces is an augmented reality experience using your phone as the camera on the world of members. Even though they invested two billion dollars in Oculus, which requires a headset, they have determined most of us will want to experience virtual worlds using our phones.

Could this concept work at your next class reunion? Instead of Poke Stops, you can use your mascot to lead alums around the campus to learn about all the new (and future) things going on.

Could your next facility tour include facts about your mission as people move from one area to another? How much food is delivered to your food bank each week, and how much you provide to people in need every month?

Your next walk-a-thon or golf tournament can include an AR component. I was part of a 5K to support our local humane society, and you picked up your race packet in the shelter. It made a difference to actually see the dogs and cats (and a few other animals). An app could have been provided giving me information about the shelter, and during the race I could have been collecting additional information and perhaps points to unlock special items at the silent auction.

At your next golf tournament there could be information about your mission at each hole to go along with yardage. While we may not all be data geeks, we all love information about new things.

Is there information in your organization that is not well known? How are you saving the dolphins? How do you help people when their house is burned down? Why is the new kind of cancer treatment working better than anything before it?

As you create your mission data inventory, think about how it can be delivered. VR and AR are just new delivery platforms. Each of your platforms will present data in its own way. Your job is to ensure they reinforce each other as well as deliver an experience crafted to take full advantage of the medium.

This is about scaling your cultivation and stewardship. Today, you most likely can only deliver deep experiences to a fraction of your donors, and prospective donors. How many people would give you more, and give to you longer, if you were able to ensure them their donations mattered?

Besides the obvious benefits of better cultivation, and stewardship, organizations also stand to gain insight into what about their mission is engaging people. Is there a part of the tour of your organization people always take extra time to explore? What are their questions? This is way beyond A/B testing of copy and images. Done right, VR and AR can become a testing laboratory.

These insights can be put to use in other channels such as social media and email. In fact, it is imperative to integrate VR and AR into your other channels. People want multiple ways of evaluating your organization. Greenpeace has a near real-time map of the deforestation and fires in Indonesia. It would not take much to give people a way to virtually tour some of those areas to see why this is so harmful.

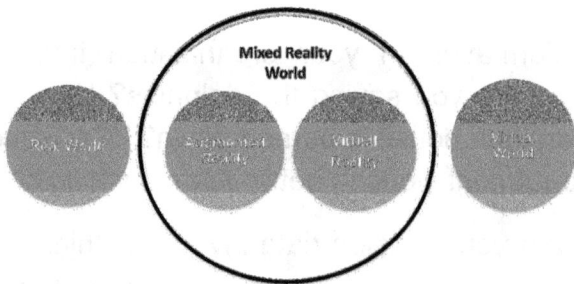

Trimble's Sketchup™3-D modeling software teamed up with Microsoft's HoloLens to create a mixed-reality platform for viewing construction projects.

https://www.sketchup.com/products/sketchup-viewer.

HoloLens is also being used at the Kennedy Space Center for Destination Mars where you can explore the red planet without the long trip, and oxygen.

https://www.jpl.nasa.gov/news/news.php?feature=6220

You might think grandma and grandpa (I go by Pops) are not going to put on a headset and visit with their grandchildren who are half-away across the country. Remember, more and more of our mature donors are the same people who can say they enjoyed rock before it became classic. I'm one of those, and I have ridden a virtual coaster, and visited Iron Man's Mansion without leaving my home. I also made it to Level 8 on Pokemon Go! thanks to my stepdaughter who we will just say made it to a higher level.

And if you just can't imagine you, or your donor, wearing a headset, then try Google Glass (take a look at the new Enterprise Edition) or wait for the AR room where you will just use your eyes. It won't be long.

Mapping Your Messages

The myriad of the ways organizations have to send their messages to constituents has resulted in message madness. Gone are the days when copy is carefully crafted and reviewed for each mailing. Turns of phrase, colors, images, and font size were debated within the organization and by consultants.

The proliferation of email turned paragraphs into sentences, and sentences into a few well-chosen words for a subject line. Social media demanded an immediacy resulting in even less control over the message.

To paraphrase McLuhan, the medium now owns the message.

Organizations need to own their messages, not the mediums. Leadership knows this, but has thrown up their hands in the face of the torrent of multi-channel outreach. There are just too many messages to manage.

With Cognitive Computing, you will need to find a new excuse.

Thanks to Natural Language Processing you can take every communication across every channel and create a true message map. You can then see how many different messages you have and how many different ways you are communicating those messages.

It will be tempting to approach this much as an editor armed with an appropriate style-guide would. I suggest waiting until you see how well messages are performing. Like Federal Express realizing we all say FedEx®, and deciding to change their name, sometimes what we think will work doesn't in the real world.

You may also discover variations on your messages work better with different constituent clusters. This is a great use of Cluster Analytics, which frees you from always looking at simple factors such as age, location, and/or gender.

Time is another factor. As you explore your data, look for changes in message performance over time. Is performance trending up or down? Don't be fooled by overall results. A message may still be yielding large donations, but is no longer resonating with particular constituents or not performing at the level it has in the past. This is why the hardest changes involve the longest-standing actions. Even in the face of declining results, we fear changing what has worked in the past.

Having these insights for your organization will also protect you from the benchmarking-trap where you make decisions based on another organization's reality instead of your own.

Take full advantage of your multi-channel outreach to continually test messaging. By taking an incremental approach you can build confidence in your insights, which will support larger message overhauls.

You can also combine Journey Analytics with message mapping to understand how your messages change (or don't change) as your relationship grows (or doesn't grow). Here are some key questions to ask:

> Do you refer to a constituent consistently across channels?

> Do you change how you refer to a constituent across channels, and is the change consistent?

> Are your ask amounts consistent (in the context of the channel) across channels?

Your constituents assume you know all of their engagement with your organization including giving history; volunteer activities; and event attendance. If you are treating someone as a prospective donor in one channel and a long-time donor in another, you risk alienating the person. Are you asking someone to volunteer who already is volunteering?

An organization with which I was working shared a horror story of one of their frontline fundraisers asking a major donor if she wanted to tour a facility where she volunteered twice a month. The donor was gracious, but also disappointed the fundraiser didn't know of her volunteer work. I'm sure the next time she was asked for money, this experience was on her mind.

Organizations can have a very long relationship with a donor. With a college, this may have started at birth as his/her parents set up a college fund aimed at an in-state school. We have all seen babies swaddled in school colors. When that baby goes to the college, graduates, gives, and then makes a planned gift, it is literally a cradle to grave relationship.

Does your messaging reflect the maturing of relationships? Use Natural Language Processing to look at your communications with long-term donors to see if you are using language reflecting the growing intimacy.

Start with looking at whether you are using their first name rather than Mr., Ms., or Mx. Donor. Do you mention children or other people in their lives? Have you visited with them and referenced those visits by mentioning the locations of those meetings?

Another way to gauge the strength of the relationship is to see if you have talked about aspects of your mission, which are not part of your general communications. This can be accomplished by comparing your communications with a particular constituent against your overall communication analysis to look for words and phrases either not found or found infrequently in overall communications.

You can use these insights to help relationship managers better understand the value of the words they use. This will also be invaluable as part of strength-of-relationship scores used to prioritize your donor pool.

We are not going to decrease the amount of communications or the ways we communicate. While there are many benefits to be gained from embracing Cognitive Computing, message insight is near the top of the list. An insight-driven organization will see all communications not as noise but as guiding signals.

Good Bots

Customer service is not a hallmark of our industry. So much of your resources are spent on delivering your mission, and raising the funds to do it, customer service can seem like something you can only deliver to your high-level donors.

The problem is donors often make a low-level gift to see how it goes. When they receive poor, or sometimes no, communications, they move on to another organization. In your own life don't you try the new restaurant for lunch before dinner? Buy something a little less expensive on a website to see how the experience goes?

The for-profit world has the same challenges, just more resources available to deal with such and even with those additional resources for-profit organizations are increasingly turning to chatbots to deliver first-level customer service.

You may well have encountered a chatbot and not realized it. That's the point. You want the experience to feel like an interaction with a human. Amazon's Echo is an example of a chatbot using audio. You ask questions, and Alexa gives you answers.

AdmitHub has created a chatbot for admissions. Prospective students can get information in a chat format whenever they want it:

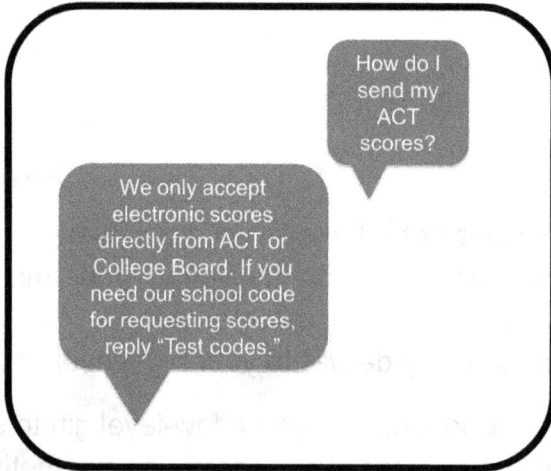

Think about how you might use this. You could provide this type of service to mission beneficiaries during hours when you don't have operators or a chatbot could provide a triage to help the majority of callers, while others are routed to live-operators. Not only will this provide 24/7/365 access, but it also may well do a better job of providing up-to-date accurate information than a live operator, especially if a plethora of possible answers exists and the information is dynamic.

Prospective donors could ask questions about planned giving or tax matters related to their giving.

Another result of implementing a chatbot is the knowledge your organization will gain regarding what visitors to your website want to know. This will help you focus on the information your constituents need rather than what you think they need.

You can also combine a chatbot with your mission impact data to provide a conversational experience as people seek to discover all the good your organization is doing. Here again you will learn deep insights into what is engaging your donors and prospective donors.

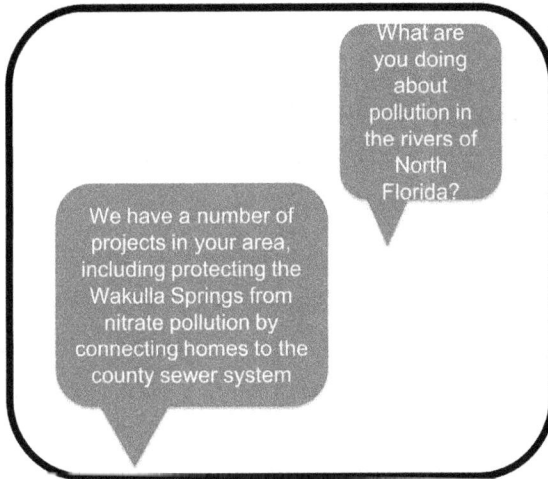

This kind of service can also be used by your staff to gain answers they need in the field or while on a call. Prospective donors, and loyal supporters, will gain confidence in your organization knowing your organization's impact is at your (or their) fingertips.

The American Cancer Society is creating a cognitive advisor for cancer patients:

> Once developed, the advisor will anticipate the needs of people with different types of cancers, at different stages of disease, and at various points in treatment. It will be dynamic and become increasingly personalized as individuals engage with it, effectively getting 'smarter' each time it is used. ACS and IBM also

envision incorporating Watson's voice recognition and natural language processing technology, enabling users to ask questions and receive audible responses. http://pressroom.cancer.org/WatsonACSLaunch

This application has a twist I believe we will encounter more and more – it gets to know *you*. This goes beyond knowing your name and location. It actually learns about your unique needs. With ACS you are talking about an incredibly sensitive topic, so you can certainly see this working with admissions where it is not just Q&A, but Question, Answer, Learn, Question, Better Answer, and so on.

More and more sectors of our economy are embracing this technology, and we can't afford to fall any further behind. We are woefully understaffed, and staff turnover is higher than it is in other sectors. This means having tools like chatbots to deliver high-level service without incurring increased costs will free-up dollars for the mission which in turn will increase fundraising as donors literally see the impact of their giving increasing in almost real-time.

Insight-Driven Relationship Management

We are in the midst of a mass migration from old (frankly ancient) database systems, many designed in the 90s, to advanced Constituent Relationship Management platforms such as Salesforce. CRM has been around for a long time, but only recently has the philanthropic community begun to let go of their clunky, yet comfortably familiar, tools.

I was a senior executive with the first company dedicated to our field to offer a true CRM – Kintera. I had sold P!N to Kintera in 2004 in large part because I believed Software-as-a-Service (SaaS) was the way the world was moving, and data was going to be a key differentiator. Before they purchased my company, Kintera was mostly known for its Friends-Asking-Friends® crowdfunding platform. You may have used it while participating in or donating to a HeartWalk® or Relay for Life®. When our P!N ProfileBuilder software was fully integrated into the Kintera Sphere® platform, it set the stage for the move to CRM as our schema already had the key components. Lori Hood Lawson, the editor of this book, oversaw the migration of ProfileBuilder and the development of the Sphere's prospect management functionality. The latter is essentially our field's version of the sales cycle.

In 2006, we introduced Social CRM (http://www.crm buyer.com/story/50769.html) to bring together relationship management; on-line fundraising; events; volunteer management; and wealth screening. The industry was intrigued, but the problem was while the platform could do amazing things it could not perform basic gift processing for donations made off-line. This was because the architecture of a SaaS multi-tenet (think apartment building) platform was not designed for transactions involving credits and debits; soft-credits; and multi-part pledge payments.

Lori and I, among many others, tried to make the case for solving the gift processing problem, but we ran out of time and hit the Great Recession precipitating Kintera being purchased by Blackbaud®. Blackbaud decided to go its own way with CRM development using .Net, and so Kintera Social CRM was relegated to processing on-line donations.

Fortunately companies, especially Salesforce, have not only kept innovating their CRM, but also focused their attention on our field. Recently, the gift processing capabilities have been dramatically improved, so CRM can finally replace donor management systems.

Salesforce.org (formerly The Salesforce Foundation), the nonprofit arm of Salesforce, gives away 10 free licenses to nonprofit organizations and also heavily discounts additional licenses. There is also an ever-growing list of vendors on the Salesforce AppExchange™ who have built applications on their platform specifically for our field.

There was a time when it was thought (including by me) CRM was going to solve all of our data problems by connecting silos on a single platform. What we did not foresee in the mid-2000s was the explosion of unstructured data and the emergence of practical Cognitive Computing. This meant CRM was never going to contain *all* of an organization's internal and relevant external data.

CRM providers came to this realization (or at least publicly stated it) and started adding data and analytic capabilities. More recently Salesforce has introduced Einstein™, its answer to Watson. Interestingly, Salesforce has a partnership with IBM to also offer Watson to its customers.

Where does a cognitive-enabled CRM fit in with the Insight Reservoir?

The Insight Reservoir is your laboratory where you discover insights worthy of action. CRM is an operational platform where you take the actions based on the insights.

One way to think of this is CRM (and other operational platforms such as marketing) need high-quality, real-time insights, while the Insight Reservoir is purposely not constrained by time or quality. This enables one of the cornerstones of Cognitive Computing – iteration.

I have a concern CRM vendors will use their cognitive capabilities to position their platforms as the single source of truth. Organizations who like a single-vendor approach (also known as "one throat to choke") to technology will be attracted to this pitch.

The best-of-vendor (BOV) vs. best-of-breed (BOB) debate has been going on for decades. BOV has won in no small part because organizations believe the integration of different applications would be smoother if they all have the same brand name. To some extent this has been true in part because vendors have made it painful to work with competitive products.

In the cognitive age integration is no longer a factor. Data can now be exchanged easily between systems without the *permission* of the system vendors. Data has been freed, so don't let vendors trick you into locking it up again.

You can now embrace BOB. Select the CRM best suited to your organization's size and needs. Do the same with other aspects of your operations from events to measuring social impact. And use your Insight Reservoir to bring all of the data together.

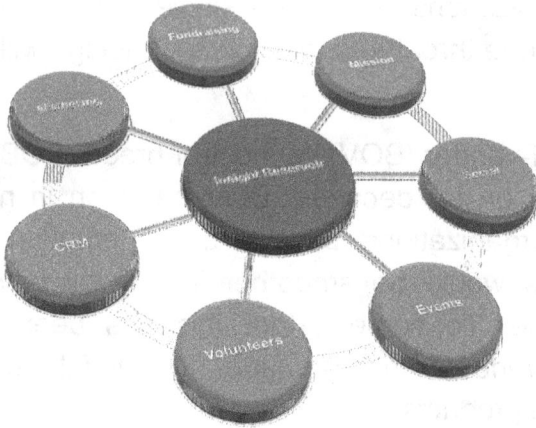

You may find a vendor who has best-of-breed solutions for more than one of your operational areas, and it's great when it happens. My point is you no longer have to feel compelled to go with one vendor to make it all work together.

What are some things you can do with a cognitive-enabled CRM?

Having real-time analysis of the relationships you are managing is number one on my list. Every time you access your CRM, you are presented with an up-to-the-minute look at the current state of every relationship. Who has paid their pledge? Who is attending the upcoming special event? Has anyone interacted with our social networks?

A score can be created based on all of the activity giving you an easy way to continually rank your pool of prospects. The Insight Reservoir enriches this in real time with insights from across your organization as well as relevant external information.

Helping you decide the best next step in a relationship is another area of opportunity. Successes and failures of different touch points can be continually analyzed and scored. It is easy to imagine a Cognitive Assistant suggesting the best project or program to focus on in your email or who might help you get a face-to-face meeting.

While the Insight Reservoir will be looking at all of this too, having real-time analysis is an invaluable tool as it brings what is happening now into the mix. The challenge will be deciding what types of actions are right for real-time analysis and which ones are not.

At the individual level, cognitive CRM makes a lot of sense. You are working closely enough with the underlying data to see when an insight is wrong. When working with groups of constituents, such as with marketing campaigns, you need to be more careful.

People may have been excluded because the time since their last gift has gone over a year in the CRM. The Insight Reservoir would have caught the reason for the change being a delay in the annual campaign mailing, and adjusted the scoring model accordingly.

Even on the individual level you need to be careful. Let's say you have a lead scoring system and it incorporates time in each stage. A prospective major gift donor with whom you have been working for the last 18 months has a family issue causing her to cancel a meeting. The cancellation triggers a lower score, and the information is sent to the head of fundraising. You spend the next couple of days calming down senior leadership.

This particular scenario is a good one to use with your major gift officers to demonstrate the importance of good notes. By putting the reason in the notes, NLP can turn "meeting cancelled because of family issue" into an insight to adjust the weight down for the cancellation.

Having an Insight-Driven CRM will soon be the norm, and the same will be true for all your other operational platforms. Successfully using your Insight Reservoir to bring all of this real-time analysis together and combine it with all of your relevant non-operational data will require imagination as much as science.

Cognitive Screening

I would be remiss if I did not address the impact of Cognitive Computing on the part of fundraising on which I have focused most of my professional time – wealth screening. Despite all of the innovation over the last 25 years, what is on the horizon will make what we have been doing seem rather quaint.

Our sector was introduced to the idea of electronically screening their donors by Marts & Lundy in 1984 when they launched the appropriately named Electronic Screening®. Consultants at M&L had seen the for-profit world use data to segment their customers, and they decided to bring it to philanthropy. Charles Headley, a pioneer in our field, was hired to program the service. A few years later Charles and I became friends after he contacted me to see how my profiling company, The Information Prospector, might use their screening service. In 1989, Charles joined Prospector, the first of many (ad)ventures we took together.

Just before Charles joined me I had found an automated solution to obtaining Securities & Exchange Commission data from The Invest/Net Group. It is hard to explain just how excited I was when I met one of the founders, John Wright, and realized we would never have to go to the SEC library to copy insider forms. Not only was this incredibly time-consuming, the SEC allowed people to re-file the forms themselves. You can imagine how well that went.

While it was great to have the data available electronically on individual stockholders, what was much more valuable was the ability to match these records to an organization's entire database. This was the birth of asset-screening. We called it Securities Prospector. Charles took it to the next level at CDA Investment Technologies which had purchased The Invest/Net Group. I joined him there in 1992.

Charles and I brought real estate into the mix in 1993. This was important because a lot more people owned real estate than were public company insiders having to report their stockholdings. When I founded P!N in 1997, I incorporated Dun & Bradstreet® data, including company ownership, and we were able to find the "Millionaires Next Door," who are mostly private company owners and executives.

Analytics started to appear in the late 90s. P!N's ProfileBuilder software had an embedded user-configurable analytics tool to score constituents based on the data the screening returned as well as giving and involvement data supplied by the client. This was an important development because the asset information was not being utilized by companies building scores for organizations. They used consumer data, and segmentation services like Prizm®, which, while powerful, lacked the depth of data at the individual level.

Today, there are many companies offering a range of data against which to screen your file. Many of them have also combined look-up services with the screening service, allowing organizations to enjoy the benefits of scale while not losing the ability to look closely at any one constituent.

Where does Cognitive Computing fit into this?

> **First**, wealth screening was Big Data before we knew what Big Data was. At P!N we had data on over 100 million households from dozens of data sources. While it was all structured, it was still Big. One of the reasons I was so quick to jump on the Big Data bandwagon was remembering how painful it was to bring all those databases together.

> **Second**, no sector has more meaningful unstructured data than ours. We have a very intimate relationship with our constituents, filled with deep dialogue as we impact people's health, education, faith, well-being, environment, society, and the future of the planet itself.

> **Third**, our sector behaves more like a sailboat than an ocean liner. The economy, social unrest, world events, natural disasters, environmental changes, and politics can stir up a wind or create a dead-calm depending on your organization's mission. Look no further than the American Red Cross and 9/11 to understand what can happen on any given day to your organization.

> **Finally**, support is driven by emotion, passion, and heart far more than logic and the size of people's bank accounts. Despite this, we have based the identification of our "best prospects" almost exclusively on wealth.

You could not ask for a more perfect fit for what Cognitive Computing can do.

Wealth screening will increasingly be seen as a part of the identification solution rather than as the solution itself. I can actually see the term *wealth screening* being rendered obsolete as wealth becomes less and less of what makes a good prospect.

The Insight Reservoir, with all of its machine and deep learning capabilities, will ingest asset and other wealth data to be part of organization-centered scores rather than vendor-centered scores. In this model you are assured any analysis will include all of the available data on a constituent rather than what you sent to the vendor and what the vendor decided was important.

Fundraisers will no longer be frustrated by a score or rating not taking into account their call report in which it was clearly stated the prospective donor recently sold their business. Gone will the days of showing up for a meeting and being asked why no one responded to their concerns voiced on your organization's social media page about the project for which you are soliciting money.

Losing control is one of the fears in the cognitive age. Actually, you are going to regain control. Too much of the screening and rating is being done in black-box systems using algorithms never fully disclosed to you. Cognitive Screening will be done in plain sight, within your Insight Reservoir, so you own the black-box and can open it whenever you want.

In what will soon be the old-school approach to screening, you send what information you have on a constituent and the information is matched to data sources by the vendor. Did you send all of your addresses (primary residence; second homes; business address)? Did you send all emails or just the one you use for outreach? With Cognitive Screening you will send all of your available information via a secure API, so you don't have to worry about creating an input file (another term on its way to the tech word museum).

One of the innovations we did at P!N was the concept of recursive matching. This is where you take what was learned from the first screening pass to increase the quality of your matches on a second pass. An example would be obtaining a business name from Dun & Bradstreet, and then using that information to match to ZoomInfo® using name and company logic.

Now imagine this happening in your Insight Reservoir. As you gain more data, you are continually matching it to both data you already have as well as sending it to screening vendors. This will be especially valuable when you perform address updating, and, of course, when you acquire new constituents.

With Cognitive Screening, scores and rating will not be static. They will change not only as the underlying data changes, but also as the algorithms become smarter over time. This will be met with excitement and trepidation. Excitement your insights are always up-to-date. Trepidation as you imagine your campaign plan being thrown into chaos as key people move in and out of your top prospect lists.

Rather than focus on your fears, keep your focus on how to best integrate dynamic information into your plans. You could have a "Trend Report" highlighting people who are moving up or down, and showing this information over time. This will help you plan for both potential good and bad news before it is too late to act.

Armed with these insights, you can develop action plans to respond to changes in your prospect pool. What are you going to do if a constituent who was rated low at the beginning of your campaign suddenly comes into money? You don't want to wait for the next campaign, so come up with the action steps to quickly integrate a new top prospect into your campaign portfolio.

Cognitive Screening does not mean wealth-screening companies will disappear. They still serve a valuable purpose by aggregating large datasets and making them available for a lot less than any one organization buying these directly. More than likely some of these companies, and new ones, will offer some version of the Insight Reservoir complete with their data. My one word of caution is to make sure you don't get vendor lock-in on the data because without the freedom to bring relevant data in from wherever that data may be located, you will greatly diminish the value of Cognitive Computing for your organization.

When you make your move to Cognitive Computing, Cognitive Screening will become inevitable for the simple reason so many other areas will be need the insights it produces. Journey Analytics will map the journey of your best prospects, and Cluster Analytics will incorporate screening data into the cluster creation process. Whether it is capacity, affinity or propensity of your constituents to give, your Cognitive Computing platform will need this information. Bringing it into your organization will ensure it is up-to-date and reflects your total relationship with each constituent.

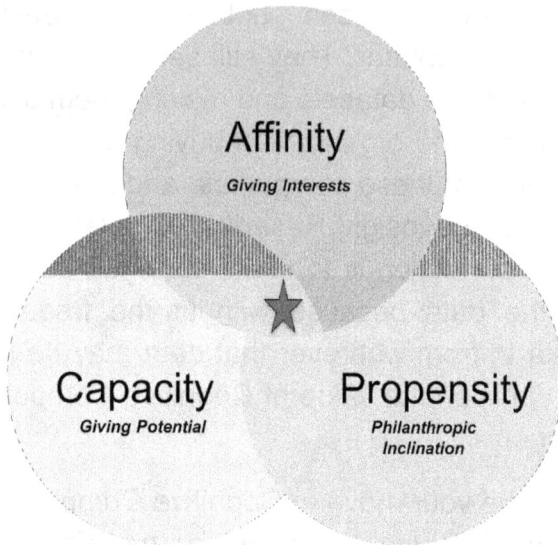

What is the Role of Prospect Researchers?

This is where I began my career. The first research I conducted was on private foundations as an assistant editor on The Foundation 500. Finding the 990-PF in the microfiche drawers at The Foundation Center in Manhattan was an introduction to the *glamour* of research. Replacing the toner in the reader-printer could have dissuaded me from this career path, yet somehow the sound of pages upon pages of grants coming out of the noisy behemoth was soothing, knowing they would lead to discoveries about the giving patterns of America's grant makers. For those of you who are researchers, you understand this love of the data journey. For those who are not, just know the people who unearth the data you need to do your job are passionate about their work.

The next summer I became the editor of The Foundation 500, and we produced six editions before I moved on to where the real money is, researching individual philanthropists. This was in the day when research profiles could be dozens of pages long, almost research novellas, as we sought to capture as much data as possible. This was all pre-web, so it was work in the library (for me and my team it was the Library of Congress), and online lookup using dial-up.

Had the highly radioactive entrepreneurial bug not bitten me, I would most likely have been a prospect researcher. Instead, I have spent my career trying to help them from the outside.

Over the years, prospect research has almost completely abandoned the old profiles for wealth screening, analytic scores, and what you can quickly find on the web. A question you may have is where does prospect research fit into a world where you bring all the internal and relevant external data together? Can you downsize or even not have a prospect research function? The answer is an emphatic *no*!

The skills a good prospect researcher possesses — data management, data curation, and the ability to work with text — are all critical to successful implementation of Cognitive Computing. When you build your ontology, you are going to want someone managing the process who knows how to organize data. When you need someone to find relevant data from a wide range of sources, you will be glad you have a professional who knows how to not only find but curate reliable sources who have ethically sourced its data.

To see the way forward you often have to look back. In the 80s, as prospect research became a profession, the skills you looked for were closely aligned with a degree in library science. At my company, The Information Prospector, a Masters in Library Science was a strong indicator you would be a good member of our team. I also liked history majors because they love to dig deep, and when I secured a stacks pass at the Library of Congress for them, they would explore every inch of this amazing national treasure.

In the 90s on-line information exploded including the World Wide Web. Being able to navigate the fee-based and free sources became a critical talent for a successful researcher. Wealth screening, including the companies I was part of or founded (CDA; Thomson; P!N), became standard parts of any large-scale fundraising campaign. This meant researchers were starting to spend more of their time analyzing, rather than gathering, data.

In the late 90s, P!N introduced ProfileBuilder™, the first software designed for researchers to find, profile, and monitor their organization's prospects. Researchers now had a way to quickly organize all of their internal and relevant external data. They also could quickly create lists of prospects based on queries they built rather than having to ask IT to create a report (which at the time was a painful process).

As the 21st century dawned, analytics started to be more than something a vendor did for you. Researchers, thanks to pioneers like Peter Wylie, were taught how they could use tools as common as Excel to mine their own data. Data science had arrived in our sector. This brought an influx of people into the profession who loved working with numbers as much as researchers loved finding a biographical article on a top donor.

The tools of the research trade had expanded dramatically, and so did the number of prospects in the portfolios of fundraisers. At the same time software was being introduced to enable the management of a relationship from prospect to gift, and from a gift to a higher gift. Today, we know it as Constituent Relationship Management (CRM).

Look no further than the name changes of the largest organization of research professionals to see how the field has evolved over the last 30 years. When I attended their first conference, APRA stood for the American Prospect Research Association. They then became the Association of Professional Researchers for Advancement. Today the profession has literally outgrown its acronym:

> For 30 years, Apra was an acronym. The "American Prospect Research Association" became the "Association of Professional Researchers for Advancement". But we are more than the sum of those letters. Our members are the professionals who drive their institutions' philanthropic missions through work in prospect development and prospect research, data analytics and data management, annual giving, advancement, special gifts and more. You can't sum it all up in four little letters.
>
> So we had to become more. More than an acronym. There is great history in our name, and a history we are proud of. So we don't want to lose that, but rather embrace it as our members continue to grow and evolve with the ever-changing field of philanthropy. http://www.aprahome.org/page/about-apra

So what is to become of the prospect researcher in the cognitive age? I see three distinct career paths: Data; Insight; and Relationship Management.

You might think data is the least valuable, but without quality data there will be no insights and no relationships to manage. The role of Chief Data Officer is where I see the people with strong library and information science backgrounds heading. Data Governance will be at the center of everything they do.

They will be responsible for what goes in the Insight Reservoir, and to whom it is delivered.

Organizations in the EU don't need to be told how important this person is. In fact, many are now looking for a Data Protection Officer (DPO) thanks to the General Data Protection Regulation (GDPR). Here is a description of the DPO from GDPR regulatory body:

- Must be appointed on the basis of professional qualities and, in particular, expert knowledge on data protection law and practices.

- May be a staff member or an external service provider.

- Contact details must be provided to the relevant Data Protection Authority (DPA).

- Must be provided with appropriate resources to carry out their tasks and maintain their expert knowledge.

- Must report directly to the highest level of management.

- Must not carry out any other tasks that could result in a conflict of interest.

http://www.eugdpr.org/key-changes.html

Note the DPO *must* report directly to senior management. No more burying data professionals in the basement of your organization. Data is one of your most valuable assets and carries with it both opportunity and risk.

Also note the DPO "must not carry out any other tasks that results in a conflict of interest." For instance, can the person who collects, curates, and governs the data also be the person who mines it for insight? This is why we need a Chief Insight Officer (CiO).

The CiO is responsible for setting up the machine and, for some organizations, deep, learning environments. The data scientists in the prospect research department will be found here. As Cognitive Computing takes hold this position will grow in importance. CiOs will need not only technical but also interpersonal skills to communicate with the people who will put their insights into action.

Divining what an organization needs to be successful; prioritizing those needs; identifying the insight implementers; and continuously evaluating the Return on Insights will be a challenging and rewarding career.

The third path is the one leading back to the core of philanthropy – relationships. The Director of Relationship Management (DRM) is the person responsible for ensuring the insights are delivered to the right people at the right time, and, most importantly, actions are taken to extract the maximum value.

In smaller organizations the CDO; CiO; and DRM may well be one person. In mid to large organizations they need to be separated to ensure each has someone accountable for outcomes. I agree with our EU friends the CDO must report to senior management. The CiO may report to IT (it's why I used a lower-case "I" because they may report to the Chief Information Officer – CIO). The DRM needs to report to the person in charge of fundraising because they will be working closely with the field officers.

Instead of being made obsolete, I see Big Data and Cognitive Computing making prospect researchers even more valuable to their organization. The only thing that may become obsolete is the term prospect researcher.

Direct Marketing – Having an Intelligent Conversation

I can remember seeing one of the first business-reply envelopes. The direct mail expert showing it was very proud as he demonstrated how with a few folds you could return a check using the same brochure asking for the money. In those days it was all about getting the most mail to the most people. Mailboxes were not bursting with solicitations, so the returns were high.

Then mail became something all organizations did, and the number of those organizations increased exponentially. This led to segmentation, where you thought as much about to whom you mailed as you did about what you were mailing them. This got direct mail results humming again.

The biggest problem is cost, and it has led to fewer organizations using the channel even though according to the Direct Marketing Association's 2015 Response Survey it outperforms digital channels, only trailing telemarketing:

> Direct mail achieves a 3.7% response rate with a house list, and a 1.0% response rate with a prospect list.

> All digital channels combined only achieve a 0.62% response rate (Mobile 0.2%; Email 0.1% for a Prospect list and 0.1% for House/Total list; Social Media 0.1%; Paid Search 0.1%; Display Advertising 0.02%).

> Telephone had the highest response rate at 9-10%

> DMA 2015 Response Survey

As technology seemingly consumes the real world, what is direct mail to do? My suggestion is to harness Cognitive Computing to re-imagine direct mail.

Rather than seeing all the other channels as competition, see them as team members. The goal is for the organization to win and not with any particular channel. With this mindset look into the insight reservoir for answers to how direct mail is interacting with the other channels and with your constituents.

Compare the tone of copy from each channel to see if there is wide variance. Determine if focus points in one is contradicting focus points in others. Consumers have very little patience for a brand without a consistent voice. Are you the scrappy innovator or the leader? Is your number one priority to cure cancer or is it to help survivors?

Once you have your copy in sync, determine if direct mail is tied into other channels. By tied, I mean are campaigns coordinated with other marketing initiatives or on their own calendar without regard to what else constituents are receiving?

A question I have asked for 25 years is, "Do you know every communication sent to every constituent from every channel for the past 12 months?" Early on no one raised his or her hand. Today, thanks to CRM, more can answer it, but when you press further, even they will tell you some channels such as social media are still outside of the communication map.

Being able to answer this important question will be a big early win for your Insight Reservoir. The answer will likely shock you. Your constituents are likely being overwhelmed with messaging, and worse, messaging that is at times contradictory.

Use the answers to build a communication map. Channel, campaign, dollars raised, date, and constituents included form the core of the map. Then layer in the messages for each. Your goal is to get the same view of your communications as your constituents have.

One concern you may have with this type of exercise is the outcome may be to cut back on a channel. While it may well be one of the outcomes, the decision should wait for an analysis of how channels are influencing each other.

The channel influence analysis is done by looking at fundraising results when campaigns happen at the same time (not necessarily the exact date, but rather a window of time), with the same messaging. What you are trying to consider is the impact of a channel (in this case direct mail) on the results in other channels. Did the mail cause me to make a donation online?

For this analysis to be accurate, you also have to take into account whether the other channels were referenced such as having the website for making an on-line donation in the copy of a direct mail piece. Ideally, you will have both this type of campaign and ones where other channels are not referenced. You can then see the difference in results.

Another key aspect of channel marketing is understanding the characteristics of the people who are responding to each channel. A concern I have about both direct mail and telemarketing results are older donors skew them. In the short term (which means as long as the older age group keeps giving), it is a big reason to keep mailing. In the long-term it means you need to build up the other channels or you are risking the organization's future.

This is where you will bring Cluster Analytics into action. Don't assume it's all about age, and generations. That's an easy out. I have helped put three children through college, and a fourth will be there in a couple of years. That means I have seen a lot of clever (and not so clever) direct mail. More importantly for those thinking direct mail is only for older donors, I have also seen my stepdaughter read the mail she found engaging.

Part of engagement has been coordinating the mail with online experiences including email and a personalized website (even one for the parent). For all my direct mail loving friends, there is hope for the future with these natives to technology.

Proper coordination of communications can lead to engagement of a 17-year-old high school student, a woman who attended high school dances when Madonna was "Like A Virgin" (no way am I saying her age in a book), plus a guy who may have hustled his way through his own teen years (music owes me!).

There is a prediction AI will write a bestseller at some point not too far down the road. Perhaps it will be about enslaved humans fighting their machine overlords. While we wait for the book, AI will first impact writing on a smaller scale – emails. Gravyty, an AI start-up, has created an application designed to create communications with donors for gift officers. Using a combination of the organization's CRM, past communications with the donor, and the writing of the gift officer, an email is drafted for the user to do final editing.

It is not hard to imagine this approach being used to create copy for direct marketing. You could study the writing of the leaders in the business, and even if you did not achieve 100% parity, you would still be far ahead of where you are now if you can't afford to employ top talent.

No area needs Cognitive Computing more than direct marketing. Without it, direct marketing will either collapse under its own multi-channel weight or be increasingly irrelevant in a world driven not by organizations, but by their constituencies.

12

The Mission is the Message

"A life is not important except in the Impact it has on other lives."

—Jackie Robinson

Having spent my career helping the fundraising side of philanthropy, I could have focused this book only on that area. I didn't for two reasons. First, I believe we need to be mission-centered in our thinking. The mission is why people give, not because you are good at fundraising. Secondly, looking at your investment in Cognitive Computing as benefiting your entire organization will make the ROi significantly higher.

The challenges organizations take on are often daunting: curing a disease; providing food for millions; and delivering clean drinking water are examples of problems needing not just money, but also extensive research. Cognitive Computing is already proving to be a great partner in these bold efforts.

Medical research is rarely a linear process. It is a series of failures, pivots, and breakthroughs. In this environment, a great deal of valuable information is hidden within research reports on all of the work being done on all of the diseases around the world.

Baylor College of Medicine was an early adopter working with IBM. Here is a synopsis of the project from IBM:

> The Knowledge Integration Toolkit reviewed and analyzed more than 310,000 scientific articles during the six-month test phase — over 300,000 more articles than the approximately 3,500 papers a highly motivated human being could have read during the same time period. What's more, the solution may have already identified correlations that would have taken humans years to uncover. The scientists at BMC anticipate that by lending unprecedented computing power to one of the greatest scientific problems of our time, the KnIT project has the potential to accelerate their work and help usher in a new era of scientific research.
>
> http://www-03.ibm.com/software/businesscasestudies/sa/en/corp?synkey=F713551X77664U56

More recently the Barrow Neurological Institute announced they had identified new genes linked to Amyotrophic Lateral Sclerosis (ALS). https://medicalxpress.com/news/2016-12-team-genes-responsible-als-imb.html

A team of graduate students at MIT and a social-service group of data scientists have come up with a way of automating parts of that evaluation process used by the SELCO Foundation to select villages in sub-Saharan Africa for a program of unrestricted cash grants to help people in low-income rural areas improve their standard of living by enabling them to buy equipment, livestock, or whatever they felt best met their needs. The system adopted by the grant-giving agency was to target the poorest villages, selected by counting the percentage of houses with thatched roofs compared with those topped by more expensive metal roofs — a task that had been carried out by fieldworkers on the ground. http://news.mit.edu/2015/satellite-imagery-aid-development-projects-0323

What projects and programs could benefit from this type of analysis at your organization? Do any of them involve a lot of research found in whitepapers; field reports; and news articles? Is there a lot of relevant data available from the government?

Mission Impact

One of the great burdens the world of social good must endure is being judged as if they were a typical for-profit business. NGOs are nonprofits for a good reason. Their business model has little hope of generating profit in the business sense of that word (I am leaving social enterprises out of this).

What entrepreneur (other than a social one) or investor (other than a philanthropic one) would want a business where you must deliver an unknown quantity of disaster related aid, to an unknown place, at an unknown time, and by the way nothing may happen at all? That's the business model for the American Red Cross. Oh, and be sure every day you are ready to bring blankets and coffee to fire victims.

How about your customers being people who may have mental illness; drug addiction; poor education; and no money? Homeless shelters have the welcome sign out every day for people with these challenges.

Despite this bad business, we judge NGOs based on their efficiency of raising funds. We beat them up when their cost of fundraising is too high, even though many of those costs are impossible to accurately allocate year to year given the long solicitation cycle for larger gifts and the impact of planned gifts.

How did we come to this? A big part of the problem is the most readily available data about NGOs is a 990 tax return, and in it is a couple of sentences about the mission and a whole lot of financial information. It makes it look like a for-profit's profit and loss statement.

I'm going to say something which may get me in a bit of trouble, but that's one of the advantages of being both passionate about philanthropy and old enough to have gone through the cycle of people telling me how wrong I am and later doing the thing they said they would never do.

Here it goes – I think the 990-H for nonprofit hospitals is a great thing for our industry, and should be adopted for all NGOs.

For those not familiar with the 990-H, it is a form nonprofit hospitals have to fill out disclosing their benefit to their communities. This means there are details about all the good these hospitals do beyond the core mission of providing care often to people who cannot pay for it.

Take a look at a couple of these reports. Every NGO should have this kind of detail in their 990. It is why they exist. Not to raise, and spend, money. It is to provide benefit to society.

Where does Cognitive Computing fit into this? If you are constantly bringing all of your impact reports into your Insight Reservoir, completing a 990-H will not be an onerous job. It will naturally flow from your work.

Now donors, and charity watchdog groups, can report on all the good your organization is doing, which puts the financial data into context. This changes the conversation with your constituents from the left-brain to the right – from finance to impact.

So how do you use the Insight Reservoir for your mission?

Start with collecting all your proposals; case statements; white papers; research reports; impact reports; field reports; and any external reviews about your mission. Next, bring in your structured data containing all the financial and beneficiary details about your projects and programs.

A good way to access this data is using a Q&A AI program like The IBM Watson Retrieve and Rank service (now part of IBM Watson Discovery). The steps involved are:

1. Collect and upload your data.
2. Train the ranking model.
3. Query the system and evaluate the results.

For your training set you are going to want a minimum of 49 questions, but to really make it work you will want at least 300 questions. There is a maximum for the training data set (300MB), but there is no maximum for the amount of data to which you will ultimately apply the trained model.

To understand this better, let's look at how they trained Watson for *Jeopardy!* They ingested Wikipedia® and MS-Encarta®, and gave it 20K *Jeopardy!* questions and answers. Don't worry, you will not need 20K questions because your organization's data will not come close to what was used to win the game.

Once you have your training set loaded, the machine learning takes over. Depending on the size of the files this might take 15-20 minutes or a few hours. When this has completed, you start querying the data and you grade the quality of the answers:

> The tool displays the first question followed by four possible answers selected from the uploaded documents. Rate each answer on a scale of one to four stars, as demonstrated in the tutorial. When you have finished, click Submit Ratings. Alternatively, you can click Do not include this question in Watson's training, I can't rate these, or Add another answer for a particular question. These alternatives are shown and discussed in the tutorial. https://www.ibm.com/watson/developercloud/doc/retrieve-rank/ranker_tooling.shtml

Every time you provide feedback, the platform is learning. All of this will create what is known as *ground truth*. Think of it as the core principles on which your data is based.

As you can imagine, the bigger the set of data and the more complex the questions, the longer it will take to learn. The good news is it doesn't forget, and it never stops learning. It is like a fusion reactor of insight.

The first deployment of your mission Q&A should be internal. This is for two reasons: 1) You can use your team as subject matter experts and editors; and 2) If there is data you don't want to be seen by the public it can be caught at this stage.

I suggest you create two versions: the first for internal use without any filtering; and the second for the public with filtering. While the multiple deployments will add some costs, you only need to train once. This approach gives staff and leadership a complete view of your information, and the public a curated view.

With the training complete, the fun begins!

For your staff, they now have all of your mission data at their fingertips. Imagine a fundraiser in the field being asked what your organization is doing about a particular issue, and with a quick natural language question the answer is there on their laptop, phone, or tablet.

Your mission team will no longer have to wonder where the latest data is or whether all the relevant data is being used for the answer.

For donors, you have an always-available impact center providing the answers they want to know about what your organization is doing with their donation. For prospective donors, you can provide a way to see the impact your organization's programs make before they give and reassure them they will have access whenever they want to your mission activities.

Another bonus is you will be able to study what your donors and prospective donors are searching. This may be at a high level where you don't know who is searching, but with some thought you can provide enough value to encourage constituents to provide their email.

Why would someone give you his or her email? Provide a monitoring service where an email is sent when new information about a project or program is entered into the platform relating to a question they have previously asked.

Don't underestimate how interested your constituents are in what your organization is doing. We are a curious species, and if you have information not available someplace else, then they are going to come to your site. And the bonus is they are very likely to share this knowledge with their friends.

One question you may have about asking questions is what if your constituents don't know how to ask a question correctly or what to ask? The good news is the system can handle natural language, so it is okay if the question has misspellings or is not a proper sentence. The other good news is you can provide "Questions You May Want to Ask" to get them started. If your mission has multiple layers, you could present them with subject areas and 5 questions next to each.

If you are still concerned about the time it will take to make all of this happen, take into consideration the value to your organization of having a deeper understanding of what your mission is accomplishing. The process of training is also teaching the teachers. No matter how much of a subject matter expert you are, this experience will deepen your understanding of your organization's impact.

Larger organizations are likely going to create their own impact centers. Small to mid-size organizations need to consider partnering with organizations focused in the same area of need to share the cost and greatly increase the value.

Even large organizations should consider joining forces with "competitors" in the interest of not only cost-savings but also the broader impact a common reservoir of domain insight could bring. In any shared environment, it will be imperative to separate truly proprietary information and information which can be shared freely.

The pay-off for all your work will be a constituency feeling much closer to the good you are doing, and a shift in the conversation from finance to impact.

Grant Makers

As I mentioned earlier, I started my career researching private foundations as Editor of The Foundation 500. If I never see another 990-PF I would be fine. Having to go through thousands of grants, categorizing each one by interest area, took a lot of time and even more patience. This experience taught me two things: figure out a way to not have to do this for the rest of my life; and there are more organizations seeking funding than there are funds.

If you are a grant maker you have opened your doors to all the need in the world. Try as you might with your guidelines, grant proposals flood in not only from organizations matching your giving interests but also from many that simply do not. One of the innovations of The Foundation 500 was to include the States of the organizations supported. Before then, people just looked at the interests of a grant maker, not considering they only gave in one or two states.

In this environment, the best answer for any incoming proposal is either a quick no or a quick yes. The worst answer is maybe. Why? Because a foundation can spend over 100 hours, getting from maybe to a final yes or no. Let's say you have two employees who can review a proposal in detail, they will be able to review 40 proposals (4,000 hours divided by 100) in detail, and that assumes they have no other assigned tasks.

This daunting math has led foundations, especially the large ones, to narrow the focus of their giving to try and stem the flow of unsolicited proposals or to just stop accepting them at all.

While I understand the thinking, this has resulted in many worthy projects and programs not receiving the funding they deserved. This reality hits newly formed organizations especially hard as well as any innovative solutions from established organizations. If you have set-up your evaluation system to look for reasons to say no, then an easy excuse is the organization, or the proposed solution, not having a track of record of success.

What if grant makers could open their doors wide again with no fear of being inundated? What if proposals could be read without any bias towards who submitted the proposal or whether the approach was new? What if staff at foundations were presented with proposals with key sections highlighted, so they can discover why a proposal should be given their precious time for further evaluation?

How would it work?

First, foundations would create Cognitive Assistants trained using proposals to which they have responded with grants and proposals which have been denied funding. Part of this includes building an ontology with emphasis on the areas of interest to them. For example, the ontology objects for The Bill and Melinda Gates Foundation would include an extensive focus on global development, health, and policy.

Next, foundation staff tests the Cognitive Assistant with funded, and unfunded, proposals that were not part of the training. Once they are comfortable, then they can ingest all new proposals this way.

If you are concerned your proposal to your go-to foundation will be lost in this cognitive shuffle, don't worry. Foundations can easily set up the system to flag known organizations (grantees).

With the growing number of foundations using an on-line proposal system, the ingestion part of the process is either already in place or could easily be added. What is needed is for a few foundations to make this investment to prove the concept.

While making proposal evaluation more efficient and effective would be reason enough to do this, there are two other reasons which may yield even more benefit for society: Identification of challenges and identification of solutions to those challenges.

Grant proposals are a window into the problems, challenges, and opportunities of communities, regions, countries, and, at times, the whole world. NGOs are societies' watchdogs, always looking and listening for problems and working on solutions.

This wealth of social intelligence has often gone unutilized because the proposals have not been seen as assets, but rather documents to be stored, or, in the case of rejections, to be discarded.

Imagine if all of the proposals for drug treatment centers were brought together. You could identify common challenges, challenges unique to an area, and challenges only found at particular centers. Now, think about the dramatic increase in heroin addiction we are witnessing.

If we had been mining proposals over the last fifteen years we would have seen this addiction-rate increase in real-time as more and more proposals highlighted the need for treatment.

Foundations seeing this trend could have adjusted their funding priorities to meet what has become a bigger killer than car accidents.

This is similar to what I learned when I spoke with an internet security company that built a new cognitive system using IBM Watson. This company found it is often low level or new activity providing the most reliable signals of where a new intrusion is either coming from now or will in the near future.

With this in mind, make sure you look not just for quantity of problems, but also for new ones. Think of your Insight Reservoir as a smoke, rather than fire, detector. The goal is to discover problems, and opportunities, when there is still time to solve or take advantage of them.

What about solutions? There are two places for us to look for those: proposals and impact reports.

Let's start with solutions. Proposals state a problem, describe a solution, and then make the case for why their organization can implement the solution. By mining the solution description, foundations can look for best-practices as well as new practices.

Proposals will yield incredible insights, but impact reports are where solutions really come alive. Fortunately, foundations (and donors in general) are demanding to know whether their money had the intended impact. We have moved way beyond a simple thank you letter, and sending an annual report for the organization.

Impact reports today detail the successes, and failures, of projects and programs. They quantify what was accomplished and include qualitative analysis. This means you have data not only on how many people are served, but also how well they were served.

256

Foundations will be able to identify what is working, and what is not, and then incorporate those insights into their evaluation of incoming proposals.

Once foundations embrace Cognitive Computing, they can form cohorts based on shared interest areas. This will dramatically increase the depth of insight, and create a community to turn those into actions. And don't forget these insights can be shared with grant seekers.

A final thought about foundations. The cost of Cognitive Computing is going down, but it is still not free. Foundations can play a role in helping NGOs benefit from this new age by supporting capacity building grants involving Cognitive Computing as well as funding cognitive projects benefiting different parts of our sector.

Private foundations are, of course, not the only sources of funds. Government funding has always helped emerging technologies find their way into our sector. The National Science Foundation has established Big Data Regional Innovation Hubs to facilitate these types of partnership between the private and public sectors. The hubs are called Spokes. The NSF put forward the following broad themes for projects:

> Accelerating progress towards addressing societal grand challenges relevant to the regional and national priority areas defined by the BD Hubs:

> > Helping automate the Big Data lifecycle; and

> > Enabling access to and spurring the use of important and valuable available data assets, including international data sets where relevant.

Here are the priority areas for the West Big Data Hub:

- Metro Data Science
- Natural Resources & Hazards
- Precision Medicine
- Data-Enabled Scientific Discovery & Learning
- Big Data Technologies

https://www.nsf.gov/funding/pgm_summ.jsp?pims_id=505264

You can see how these priorities line up very well with NGO's missions. Visit the hub for your region to discover their current priorities, and also look at their awards and partners.

Millions of dollars in funding are available, but they require organizations to think creatively and to also form relevant partnerships. Not surprisingly, university research offices are aggressively pursuing this funding. I would like to see NGOs equally engaged both as direct solicitors of funds and also as partners with academia and/or private industry.

13

Data Governance – De-risking Your Organization

"I have as much privacy as a goldfish in a bowl."

—Princess Margaret

Data Governance is not a new topic. In fact, over the years there has been plenty of talk (and even concrete actions) about data quality and security, but Big Data has put a spotlight on the lack of true Data Governance.

A sign this is changing is how the term Data Governance has morphed to be Information Governance at many organizations. This is a recognition data is only the first of three layers:

Data: The raw data elements.

Information: A collection of data elements, and the results of analysis of those elements.

Insight: The meaning derived from the information, and the decisions and actions taken based on it (Actionable Insights).

All three layers must have governance as each creates its own risks. Since Insight is the highest level, we may see Insight Governance being the ultimate name for this, but for now Data/Information Governance is the nomenclature.

One of the fears of Big Data is what you might discover if you bring all of your databases together. We grew up with a "garbage in = garbage out" mindset, which made us reluctant to bring information into a database without a lot of oversight.

In the Big Data age, we need to think "data in = insight out," where we recognize data is imperfect, but we are not concerned because the Insight Reservoir you have established is designed with those imperfections in mind.

Once all of your data is in the reservoir you will have a complete picture of all the information you have across your organization. You can create single golden records for constituents enabling you to not over communicate and, more importantly, communicate with the full knowledge of their relationship with you.

Data quality goes up because you no longer have a good address in one data silo and a bad one in another. You can also clean up your data, so names are correctly spelled and formatting is correct. You can also identify where you have data that should not be in your database, such as social security numbers or credit card numbers.

It may seem like an exercise designed for data and IT professionals, an exercise leadership can sit out. In the corporate world we are seeing Data Governance led by a Chief Data Officer reporting to the CEO. We need to follow their lead.

For leadership, Data Governance is about managing risk. Hospital fundraisers were the first to encounter data as a risk with the Health Insurance Portability and Accountability Act of 1996 (HIPAA). This was the first time regulations were put in place to govern the use of data by fundraisers. Hospital foundations no longer had access to certain patient information such as diagnosis and primary physician.

Prior to HIPAA, hospital foundations had begun using wealth screening to identify patients who had high giving potential. HIPAA stopped this activity in its tracks for a while as hospitals' legal counsel tried to figure out what was and what was not allowed.

A strong Data Governance program would have enabled hospitals to quickly determine what information their foundations had; delete data in violation of HIPAA; and implement processes to ensure compliance moving forward. Instead, hospitals wasted valuable time wringing their hands not knowing whether they were in compliance or not.

More recently we have seen the European Union enact The EU General Data Protection Regulation (GDPR). This is the most expansive data privacy regulation ever implemented. While readers who are not in the EU may think this has nothing to do with them, think again. These regulations extend to any EU citizen in your database.

> "Arguably the biggest change to the regulatory landscape of data privacy comes with the extended jurisdiction of the GDPR, as it applies to all companies processing the personal data of data subjects residing in the Union, regardless of the company's location."
> http://www.eugdpr.org/key-changes.html

The GDPR grew out of regulators hearing from constituents who felt they were being overwhelmed with solicitations, many from organizations which did not interest them. This was caused in part by list sharing, a common practice in the direct mail business. It was also a result of better data appending services able to provide email addresses and other information such as wealth indicators.

The privacy rights are very broad, and the impact on fundraising in the EU will be profound. The clock will be turned back decades to a time when fundraising used a scatter shot approach hoping to get a response from someone. The EU is asking organizations to continue doing their good work, but now without any modern tools. The result will be lower revenue, higher costs, and far less money to deliver their mission.

My prediction is GDPR will be modified down the road (as HIPAA regulations have now been) as organizations are forced to cut back, or even end, operations. Ultimately people will realize all of the data was being collected not for a nefarious reason, but in order to steward their funds more efficiently.

This does not mean privacy is unimportant. There are many aspects of GDPR that are good, and all of its intentions have value. The problems start with a lack of understanding about why data is collected in the first place. Making it illegal to store information about constituents without their consent means organizations will not have the data necessary to properly segment their constituency.

Imagine for a moment Amazon was unable to show you what you are interested in, but rather was like the old Sears catalog with only sections (electronics; toys; hardware) as a way to find what you are looking for. You would grow frustrated quickly as you clicked from page to page. You would also be very upset to find you paid more for the product than you needed to because you didn't get to the lowest cost provider.

The only way to gather information will be to gain the consent of the constituent. This may sound noble, but in practice will be a nightmare. My suggestion is EU organizations need to get out in front and fully explain why they do this. When people realize the services they count on are directly impacted by the efficiency of the fundraising process, they will be more likely to grant the consent.

> "The conditions for consent have been strengthened, and companies will no longer be able to use long illegible terms and conditions full of legalese, as the request for consent must be given in an intelligible and easily accessible form, with the purpose for data processing attached to that consent. Consent must be clear and distinguishable from other matters and provided in an intelligible and easily accessible form, using clear and plain language. It must be as easy to withdraw consent, as it is to give it."
> http://www.eugdpr.org/key-changes.html

Here are the rights given to every EU citizen regardless of where the organization is headquartered:

Breach Notification

Under the GDPR, breach notification will become mandatory in all member states where a data breach is likely to "result in a risk for the rights and freedoms of individuals." This must be done within 72 hours of first having become aware of the breach. Data processors will also be required to notify their customers, the controllers, "without undue delay" after first becoming aware of a data breach.

Right to Access

Part of the expanded rights of data subjects outlined by the GDPR is the right for data subjects to obtain from the data controller confirmation as to whether or not personal data concerning them is being processed, where and for what purpose. Further, the controller shall provide a copy of the personal data, free of charge, in an electronic format. This change is a dramatic shift to data transparency and empowerment of data subjects.

Right to be Forgotten

Also known as Data Erasure, the right to be forgotten entitles the data subject to have the data controller erase his/her personal data, cease further dissemination of the data, and potentially have third parties halt processing of the data. The conditions for erasure, as outlined in article 17, include the data no longer being relevant to original purposes for processing, or a data subjects withdrawing consent. It should also be noted that this right requires controllers to compare the subjects' rights to "the public interest in the availability of the data" when considering such requests.

Data Portability

GDPR introduces data portability - the right for a data subject to receive the personal data concerning them, which they have previously provided in a *'commonly used and machine readable format'* and have the right to transmit that data to another controller.

Privacy by Design

Privacy by design as a concept has existed for years now, but it is only just becoming part of a legal requirement with the GDPR. At its core, privacy by design calls for the inclusion of data protection from the onset of the designing of systems, rather than an addition. More specifically - *'The controller shall...implement appropriate technical and organisational measures...in an effective way...in order to meet the requirements of this Regulation and protect the rights of data subjects'.* Article 23 calls for controllers to hold and process only the data absolutely necessary for the completion of its duties (data minimisation), as well as limiting the access to personal data to those needing to act out the processing.

Data Protection Officers

Currently, controllers are required to notify their data processing activities with local DPAs, which, for multinationals, can be a bureaucratic nightmare with most Member States having different notification requirements. Under GDPR it will not be necessary to submit notifications / registrations to each local DPA of data processing activities, nor will it be a requirement to notify / obtain approval for transfers based on the Model Contract Clauses (MCCs). Instead, there will be internal record keeping requirements, as further explained below, and DPO appointment will be mandatory only for those controllers and processors whose core activities consist of processing operations which require regular and systematic monitoring of data subjects on a large scale or of special categories of data or data relating to criminal convictions and offences. Importantly, the DPO:

- Must be appointed on the basis of professional qualities and, in particular, expert knowledge on data protection law and practices

- May be a staff member or an external service provider

- Contact details must be provided to the relevant DPA

- Must be provided with appropriate resources to carry out their tasks and maintain their expert knowledge

- Must report directly to the highest level of management

- Must not carry out any other tasks that could results in a conflict of interest.

To learn about the GDPR, and keep on top of changes visit http://www.eugdpr.org/.

The fallout from these regulations, which do not go into effect until May 25, 2018, has already begun. Wealth screening companies have stopped offering services to EU organizations; technology providers have modified their platforms to enable the notification provisions; direct marketing firms are adjusting their business models; and organizations are implementing processes, and hiring staff, to manage compliance.

When HIPAA came along, non-hospital organizations in the U.S. looked at it as somebody else's problem rather than a wake-up call to get serious about data governance. We cannot do the same with the GDPR.

Our friends in the EU don't have a choice. I recommend you imagine your organization faced with the same regulations as you design your Data Governance program. Hopefully the draconian measures in the GDPR will not spread, but you must be ready.

Fortunately organizations like the Association of Records Managers and Administrators (ARMA International) have been serious about Data Governance long before GDPR came along. ARMA has established principles; processes; and certifications for Data Governance.

Let's start with the principles:

ARMA International has established Generally Accepted Recordkeeping® Principles for Data Governance:

Accountability A senior executive (or person of comparable authority) shall oversee the information governance program and delegate responsibility for records and information management to appropriate individuals. The organization adopts policies and procedures to guide personnel and ensure that the program can be audited.

Transparency An organization's business processes and activities, including its information governance program, shall be documented in an open and verifiable manner, and the documentation shall be available to all personnel and appropriate interested parties.

Integrity An information governance program shall be constructed so the information generated by or managed for the organization has a reasonable and suitable guarantee of authenticity and reliability.

Protection An information governance program shall be constructed to ensure a reasonable level of protection to records and information that are private, confidential, privileged, secret, classified, essential to business continuity, or that otherwise require protection.

Compliance An information governance program shall be constructed to comply with applicable laws and other binding authorities, as well as with the organization's policies.

Availability An organization shall maintain records and information in a manner that ensures timely, efficient, and accurate retrieval of needed information.

Retention An organization shall maintain its records and information for an appropriate time, taking into account its legal, regulatory, fiscal, operational, and historical requirements.

Disposition An organization shall provide secure and appropriate disposition for records and information that are no longer required to be maintained by applicable laws and the organization's policies.

Generally Accepted Recordkeeping Principles® ©2017 ARMA International, www.arma.org/principles

In addition, ARMA has an Information Governance Maturity Model with five levels:

Level 1 (Substandard): This level describes an environment where information governance concerns are not addressed at all, are addressed minimally, or are addressed in a sporadic manner. Organizations at this level usually have concerns that the information governance programs will not meet legal or regulatory requirements and may not effectively serve their business needs.

Level 2 (In Development): This level describes an environment where there is a developing recognition that information governance has an impact on the organization and that the organization may benefit from a more defined information governance program. The organization is vulnerable to redress of its legal, regulatory, and business requirements because its practices are ill-defined, incomplete, nascent, or marginally effective.

Level 3 (Essential): This level describes the essential or minimum requirements that must be addressed to meet the organization's legal, regulatory, and business requirements. Level 3 is characterized by defined policies and procedures and the implementation of processes specifically intended to improve information governance. Level 3 organizations may be missing significant opportunities for streamlining the business and controlling costs, but they demonstrate the key components of a sound program and may be minimally compliant with legal, operational, and other responsibilities.

Level 4 (Proactive): This level describes an organization-wide, proactive information governance program with mechanisms for continuous improvement. Information governance issues and considerations are routinized and integrated into business decisions. For the most part, the organization is compliant with industry best practices and meets its legal and regulatory requirements. Level 4 organizations can pursue the additional business benefits they could attain by increasing information asset availability, as appropriate; mining information assets for a better understanding of client and customer needs; and fostering their organizations' optimal use of information assets.

Level 5 (Transformational): This level describes an

organization that has integrated information governance into its infrastructure and business processes such that compliance with the organization's policies and legal/regulatory responsibilities is routine. The organization recognizes that effective information governance plays a critical role in cost containment, competitive advantage, and client service. It implements strategies and tools for ongoing success.

Information Governance Maturity Model ©2017 ARMA International, http://www.arma.org/r2/generally-accept ed-br-recordkeeping-principles/information-governance-maturity-model

As you form (or improve) your organization's Data Governance, note senior management's involvement is what gets you to level 5. While you need a manager for Data Governance, without senior management it is unlikely you will be able to break through the silos and obtain the resources needed to implement an effective program.

All of this activity centers on privacy. The best way to ensure privacy is to have a strong Data Governance program. It is not possible to decide what you can and cannot store if you don't know everything you have. There is no way to know who has access to what information unless you bring it all together and study usage as well as sharing.

As I said earlier, one of the reasons I like Insight Reservoir instead of Data Lake as the way to describe where all your data is stored, managed, and transformed into insights is it implies it is something *owned* by you. Who has access to your reservoir and how much of it they can have are determined by you.

You can make decisions to anonymize some data and allow limited access to other data. Leadership has the cross-organizational view, while others have a more limited, but still valuable, view for their uses.

Here are some initial actions steps to take if you have not already implemented a Data Governance program:

- Establish a Data Governance Committee with a representative from each of the major departments within your organization.
- Identify all internal data sources and relevant external data sources you are using.
- As you identify the data sources, also identify who has access to the data.
- Identify actions being taken based on insights to ensure those actions are in line with organization and governmental policies.

The last item highlights an area which has seen the least amount of governance, the use of insights. Insight users have taken the position the insight creators do not govern their actions. This is not how regulators and the public see it.

The EU GDPR came about in part because of too many mailings being received by people, from organizations they never supported. This was the result of these organizations being part of list cooperatives where one can mail to other organizations' donors. Organizations were also appending data from external sources to help them segment their constituents based on demographic and economic data.

College admissions' offices have been in the news regarding their practice of looking at the giving potential of parents when deciding if applicants should be admitted. The practice of evaluating incoming parents is common, but using the information to potentially admit an applicant based on wealth over another who might be academically more qualified is something the public did not know. What made this all the worse was the policy of the university was to be *need-blind* which implies the process is not influenced by finances.

What this shows is the transformation of data into actionable insights must be governed end-to-end if an organization is to fully understand its risks. Only this approach will ensure your insights are used in accordance with your organizations policies and those of governmental regulators.

If you are seriously considering a Big Data project, it means you are already a Big Data organization. You just haven't implemented the technology to manage it. This means right now you have ungoverned Big Data, and that is what you should be concerned about.

CHAPTER

14

Should You Fear the Future?

"If you want to conquer fear, don't sit home and think about it.
Go out and get busy."

—Dale Carnegie

Fear may seem like an odd chapter to have in a book about technology and data. Frankly, I believe all books about technology should have one. Whether it is the basic fear of change; fear of the new technology not working; fear you will lose your currently beloved technology; fear you won't know how to use the new technology; or fear your investment in the new technology will be wasted, fear is pretty common in this area.

With Cognitive Computing there are two added fears: the fear of being replaced and the fear it might take over the world. I will not address the classic fears which many change management books have taken on. My focus will mostly be on the last two, but I will share some thoughts about the classics:

Fear of Change: Cognitive Computing and Big Data are here to stay, so rather than fight the future continue doing what you are doing right now – understand what it is; what it can do; and how you can put it to work for you and your organization.

Fear of Losing Your Currently Beloved Technology: Let's face it, philanthropy has a lot of very old and tired tech. We keep trying to put a pretty front-end on it, but ultimately it lacks the ability to engage with our constituents. It's time to move on.

Fear You Won't Know How to Use It: See *Fear of Change*, and take advantage of the amazing amount of free learning experiences only a click away, some of which you will find in the *So You Want to Know More* chapter at the end of this book.

Fear Your Investment Will Be Wasted: This is the most legitimate of all the fears, but not for the reason you might think. The technology typically works. What doesn't is the organization's willingness to adapt processes, procedures, and culture to utilize it. Cognitive Computing and Big Data are transforming industries from healthcare to finance, so it definitely works. Focus this fear on your organization, making sure it is ready (and you are) to use it.

Now on to the Big Two Fears.

The fear of being replaced is as old as the first machines, but a machine was designed to replace your muscles. Cognitive Computing has the potential to replace your brain. Until recently this was just the stuff of science fiction movies, and at the very least was not going to happen in your lifetime.

Then you read headlines like "Japanese company replaces office workers with artificial intelligence."

This involved 34 employees who calculated the payout to healthcare providers based on claims submissions. Interestingly, no pay-outs will be made without a human double-checking the AI's math.

When you look deeper into what the employees did, it involved a lot of calculations. This is the type of job that will either be replaced or reimagined as the technology takes hold. In the world of philanthropy this could involve people who are calculating estate planning pay-outs; delivering insights from giving data; determining staffing levels based on mission demand; inventory related to the mission (American Red Cross coffee, for instance).

What is not going to be replaced are the people who are making strategic decisions. In fact they will be given Cognitive Assistants to help them make better decisions, faster. It will also not replace people who need to continually use their judgment to decide between different alternatives, and those decisions are not ones where a probability of success is enough.

Your Cognitive Assistant might provide you a list of people to invite to your event, which has a maximum capacity, and while there are people who are on the list for sound reasons, you still invite others because you know something the Cognitive Assistant doesn't or in your judgment it is the best thing to do for your organization.

As you think about your relevance AKA employment, think about how much of your job relies on your judgment and how much of your job involves making calculations or retrieving data.

Also keep in mind, the philanthropic world is understaffed, overworked, and underpaid. Ours is a field where having Cognitive Assistants may be the difference between getting a job done or not.

The Center for the Future of Work (www.futureofwork.com) is a great source of thought leadership on the topic of what work might look like as AI becomes fully integrated into society. They strike an optimistic tone:

> "The future that we have sketched out in *What to Do When Machines Do Everything* is a future in which healthcare is better, education is better, government services are better, and wealth is more evenly distributed. We believe that in another generation we will look back at 2017 and laugh (or cry) at how bad many things were and how much we put up with. The 'great digital build out' that is accelerating – powered by AI and next generation smart 'things' – is going to see us build a future that is, by any objective standard, better than the world of today. In doing this, the work that we do – and yes, there will still be plenty of work for us to do – will be better, more enjoyable, and more lucrative than work has ever been."
> http://www.futureofwork.com/article/details/the-future-of-work-is-a-rorschach-test-what-do-you-see

Here are some questions to ask as you ponder your future job security:

- Does my job require me to make frequent judgments about next steps?
  - Are those judgments subjective requiring experience or do they involve pre-set variables easily learned?

- Am I deciding how much toothpaste to ship to a store vs what sentence to give a convicted criminal?
  - Are the consequences of a wrong decision high?
    - Surgeon vs a barista
    - Death vs a bad cup of coffee

- Is human interaction a necessary aspect of your job?
  - Major gift officer meeting with a prospect vs sending a solicitation email
  - Does the human interaction involve more than acquiring information?
    - Does the human interaction involve forming a relationship?

- Is your human analysis integral to the delivery of your work product?
  - Do you provide insights which can be acted on?

- Does your job require you to act on the insights of other people?

o Does your job require you to act on the insights of computers and/or machines?

o What is the value of your job to your organization?

o What is the cost of your job to your organization?

Our first thought is naturally about losing today's job, but what about the new jobs that will be created? No one foresaw the need for webmasters or social media managers even though we take those positions for granted now.

At this point we are just beginning to see cognitive jobs appear. Here are some I believe will be commonplace in the near future:

- Cognitive Teachers
- Cognitive Ethicists
- Cognitive Team Managers
- Cognitive Curriculum Developers
- Journey Designers
- Journey Monitors
- Insight Action Designers
- Insight Curators
- Cognitive Insight Translators
- IoT Application Designers
- IoT Field Maintenance
- Augmented Reality Designers
- Augmented Reality Event Managers

This is in addition to the exploding demand for data scientists. In my service on the Advisory Board of the University of Central Florida's Master of Science in Data Analytics, I have seen firsthand both the interest in learning data science and the insatiable demand for those skills. Cognitive expertise is starting to see the same demand curve, and it will surely accelerate as adoption increases.

Take a look at computer and mathematical science related occupations in 1997 and in 2016 as reported by the Bureau of Labor Statistics. What you see are two things: first, the number of jobs has increased dramatically; and second, the diversity of jobs has also increased.

Bureau of Labor Statistics

1997	Total Employment	2016	Total Employment
Systems Analysts, Electronic Data Processing	530,420	Computer and Information Research Scientists	26,580
Data Base Administrators	82,600	Computer and Information Analysts	665,830
Computer Support Specialists	406,230	Computer Systems Analysts	568,960
Computer Programmers	501,390	Information Security Analysts	96,870
Computer Programmer Aides	63,240	Software Developers and Programmers	1,604,570
Programmers, Numerical Tool and Process Control	8,500	Computer Programmers	271,200
All Other Computer Scientists	82,630	Software Developers, Applications	794,000
Operations and Systems Researchers and Analysts, Except Computer	71,530	Software Developers, Systems Software	409,820
Mathematical Scientists	8,280	Web Developers	129,540
		Database and Systems Administrators and Network Architects	647,610
Statisticians	15,090		
Actuaries	11,770	Database Administrators	113,730
Financial Analysts, Statistical	43,930	Network and Computer Systems Administrators	376,820
All Other Mathematical Scientists	5,190	Computer Network Architects	157,070
Mathematical Technicians	1,590	Computer Support Specialists	791,580
TOTAL	**1,832,390**	Computer User Support Specialists	602,840
		Computer Network Support Specialists	188,740
		Miscellaneous Computer Occupations	261,210
		Computer Occupations, All Other	261,210
		Mathematical Science Occupations	167,770
		Actuaries	19,940
		Mathematicians	2,730
		Operations Research Analysts	109,150
		Statisticians	33,440
		Miscellaneous Mathematical Science Occupations	2,510
		Mathematical Technicians	510
		Mathematical Science Occupations, All Other	2,000
		TOTAL	**8,306,230**

Also note the BLS does not use the term Data Scientists. Instead they use Computer and Information Research Scientist. How long will it take them to use the word cognitive or artificial intelligence?

So far we have looked at individual fears, but what about the organization? NGOs have traditionally not been nearly as concerned about being replaced as their for-profit brethren are. For so long it seemed organizations would not only survive but more and more of them would be created each year.

Today it is no longer uncommon to hear about organizations either shutting down or merging. Even higher education is not immune to the realities of the marketplace. Healthcare has been undergoing consolidation for years, and there is more to come.

Add the disruptive nature of Cognitive Computing to the mix, and leadership needs to take their thinking to a high enough level to see not only the potential changes for employees, and mission beneficiaries, but also to how the organization operates.

Earlier I focused on Sears and Amazon to illustrate the difference between an inward focused organization and an insight driven one. Now let's look at an organization even older than Sears – the United States Postal Service.

It was not that long ago many people wondered if there was a future for the USPS. After all, hadn't email replaced letters? Online advertising meant fewer catalogs, and direct mail volume dropped as organizations turned to other marketing channels.

Rather than do what you might expect a government agency to do – keep your head firmly planted in the past – the USPS saw an opportunity emerging from the very technology which had seemed to be its demise – the internet.

This is from the USPS 2016 Annual Report:

> Our Shipping and Packages revenue continues to show strong year-over-year growth as a result of our successful efforts to compete in the shipping services and the "last-mile" e-commerce fulfillment markets, including through Sunday deliveries.
>
> Volume has also experienced end-to-end growth as we have responded to customers' increased use of online shopping, which provided a surge in package volume with a record number of packages delivered during the calendar year 2015 holiday season.
>
> To accommodate the surge in volume and to minimize service disruptions during the holiday season, we have increased Sunday delivery service and added non-career employees for the holiday season in accordance with our labor agreements.

Cognitive Computing is surely going to disrupt every industry in some way, and your organization will not be immune. What is critical is for you to focus on what is now possible, rather than on what is no longer viable. If the Post Office can do it, then certainly no organization on the planet can claim they are too old or too bureaucratic to change.

You might think companies seeking to benefit from Cognitive Computing would take a "damn the torpedoes" (and the planet) approach, but fortunately this is not the case. Here are IBM's three core principles for the development of AI:

Purpose: Technology, products, services and policies should be designed to enhance and extend human capability, expertise and potential. They should be intended to augment human intelligence, not replace it.

Transparency: AI systems will make clear when and for what purpose it is deployed and all major sources of data that inform its solutions.

Opportunity: Developers of AI applications should accept the responsibility of enabling students, workers and citizens to take advantage of every opportunity in the new economy powered by cognitive systems. They should help them acquire the skills and knowledge to engage safely, securely and effectively in a relationship with cognitive systems, and to perform the new kinds of work and jobs that will emerge in a cognitive economy.

Will AI take over the world? I say yes, but not in the iRobots™ or Bladerunner™ version. I see AI much more like electricity. It will power innovations enabling us to solve what seemed like unsolvable problems, including climate change, cancer, hunger, clean-water, and caring for the growing elderly population. Those innovations will in turn create employment, replacing the jobs lost to AI, at least in part.

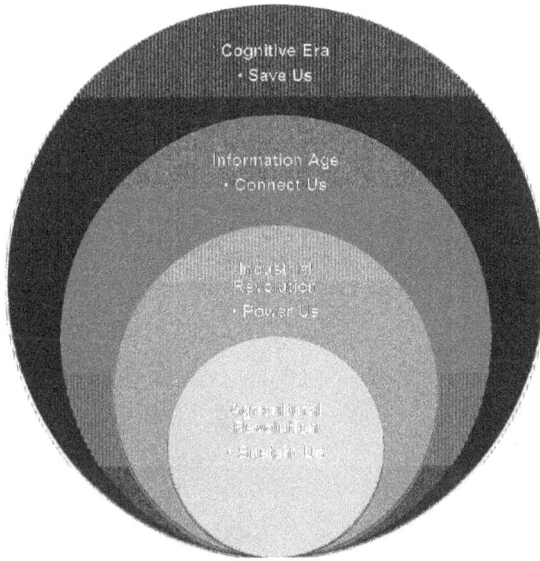

SHOULD YOU FEAR THE FUTURE?

Cognitive Era
· Save Us

Information Age
· Connect Us

Industrial
Revolution
· Power Us

Agricultural
Revolution
· Enrich Us

285

15

Take the Future for a Spin

"The future is but the present a little farther on."

—Jules Verne

One of my goals in writing this book was to help people understand Cognitive Computing is here and ready to be put to good use. Though it sounds futuristic, it is already available for everyone to utilize no matter what size your organization is. All we are missing is the willingness to try something new and the imagination to use it to its full potential.

In this spirit, I have put together sites where you can try out the different aspects of Cognitive Computing we have covered. All of them are free (or at least free as I write this) to test. Start with the examples they have pre-installed, but whenever possible put some of your own data into the system:

- Marketing copy from your latest campaign.
- Emails you have written, and ones you have received.
- Pictures of your organization's logo and any locations unique to your mission such as your headquarters.
- Your campaign video.
- A funding proposal.
- Audio from your call center or a recorded presentation by you or a member of your team.

And don't forget to play. You are not going to break anything. You are not going to unleash human-hating robots. Bring the joy of discovery you had as a child. There will be plenty of time to be a grown up when you embark on your cognitive journey.

Let's start with the first star of the cognitive age, IBM Watson. IBM has enabled all of us mere mortals to access each of the major functions using APIs. When I designed one of the first Watson commercial applications, we had access only to the Q&A functionality used to win *Jeopardy!*. It is a lot easier to work with Watson today than it was in 2014.

The services are offered on the IBM Bluemix platform:

> **Visual Recognition** uses deep learning algorithms to analyze images that can give you insights into your visual content. You can organize image libraries, understand an individual image, and create custom classifiers for specific results that are tailored to your needs.
>
> DEMO: https://visual-recognition-demo.mybluemix.net/
>
> TRAIN: https://visual-recognition-demo.ng.bluemix.net/train

Speech to Text service uses speech recognition capabilities to convert Arabic, English, Spanish, French, Brazilian Portuguese, Japanese, and Mandarin speech into text.

DEMO: https://speech-to-text-demo.mybluemix.net/

Conversation service allows you to understand what users are saying and respond with natural language.

DEMO: https://conversation-demo.mybluemix.net/

Natural Language Classifier applies deep learning techniques to make predictions about the best predefined classes for short sentences or phrases.

DEMO: https://natural-language-classifier-demo.myblue mix.net/

Natural Language Understanding is a collection of APIs that offer text analysis through natural language processing. This set of APIs can analyze text to help you understand its concepts, entities, keywords, sentiment, and more. Additionally, you can create a custom model for some APIs to get specific results that are tailored to your domain.

DEMO: https://natural-language-understanding-demo. mybluemix.net/

Discovery adds a cognitive search and content analytics engine to applications to identify patterns, trends and actionable insights that drive better decision-making.

DEMO: https://discovery-news-demo.mybluemix.net/

Language Translator service converts text input in one language into a destination language for the end user using background from domain-specific models. Translation is available among Arabic, Chinese, English, French, Portuguese, German, and Spanish (some languages may not be available for all domains).

DEMO: https://language-translator-demo.mybluemix. net/

Personality Insights extracts personality character-istics based on how a person writes. You can use the service to match individuals to other individuals, opportunities, and products, or tailor their experience with personalized messaging and recommendations. Characteristics include the Big 5 Personality Traits, Values, and Needs. At least 1,200 words of input text are recommended when using this service.

DEMO: https://personality-insights-livedemo.myblue mix.net/

Text to Speech converts written text into natural sounding audio in a variety of languages and voices. You can customize and control the pronunciation of specific words to deliver a seamless voice interaction that caters to your audience. Use text to speech to develop interactive toys for children, automate call center interactions, and communicate directions hands-free.

DEMO: https://text-to-speech-demo.mybluemix.net/

Tone Analyzer uses linguistic analysis to detect three types of tones in written text: emotions, social tendencies, and writing style. Use the Tone Analyzer service to understand emotional context of conversations and communications. Use this insight to respond in an appropriate manner.

DEMO: https://tone-analyzer-demo.mybluemix.net/

Tradeoff Analytics applies decision analytics technology, enabling users to avoid choice overload when making complex decisions involving multiple, conflicting goals.

DEMO: http://tradeoff-analytics-demo.mybluemix.net/

As you go through the Watson demos, take a moment to enjoy how amazing it is we have access to billions of dollars of research and development at no cost, and for low transactional costs we can put it to work for good.

Watson may have been the first star, but he is not the only cognitive platform. All of the tech titans from Microsoft to Google to Amazon are racing to build their own. Salesforce, recognizing CRM is not going to solve all of an organization's data problems, has introduced Einstein.

The Microsoft Azure platform is where you will find Microsoft's cognitive applications. While Microsoft may be a little late to the party, they are catching up quickly, and in some instances are surpassing the competition.

Computer Vision API

- Returns information about visual content found in an image. Use tagging, descriptions, and domain-specific models to identify content and label it with confidence. Apply the adult/racy settings to enable automated restriction of adult content. Identify image types and color schemes in pictures.

- Read text from images using optical character recognition (OCR) to extract the recognized words into a machine-readable character stream. Analyze images to detect embedded text, generate character streams, and enable searching.
- Read handwritten text from images using handwritten OCR. It works with different surfaces and backgrounds, such as white paper, yellow sticky notes, and whiteboards.
- The Celebrity and Landmark Models are examples of Domain Specific Models. Their celebrity recognition model recognizes 200K celebrities from business, politics, sports and entertainment. Their landmark recognition model recognizes 9,000 natural and manmade landmarks from around the world.
- Analyze video in near real-time extracting frames of the video from your device and then sending those frames to the API calls of your choice.
- Generate a high quality storage-efficient thumbnail based on any input image. Use thumbnail generation to modify images to best suit your needs for size, shape, and style. Apply smart cropping to generate thumbnails that differ from the aspect ratio of your original image, yet preserve the region of interest.

DEMO: https://azure.microsoft.com/en-us/services/cognitive-services/computer-vision/

Video API Intelligent video processing produces stable video output, detects motion, creates intelligent thumbnails, and detects and tracks faces.

DEMO: https://azure.microsoft.com/en-us/services/cog nitive-services/video-api/

Video Indexer builds upon media AI technologies to make it easier to extract insights from videos. Search for spoken words, faces, characters, and emotions.

DEMO (free trial): https://vi.microsoft.com/

Content Moderator uses machine-assisted moderation of text and images, augmented with human review tools.

DEMO (free test drive): https://contentmoderator. cognitive.microsoft.com/

Face API Detect human faces and compare similar ones; organize images into groups based on similarity; and identify previously tagged people in images.

DEMO: https://azure.microsoft.com/en-us/services/cog nitive-services/face/

Emotion API takes a facial expression in an image as an input, and returns the confidence across a set of emotions for each face in the image, as well as bounding box for the face, using the Face API. If a user has already called the Face API, they can submit the face rectangle as an optional input. The emotions detected are anger, contempt, disgust, fear, happiness, neutral, sadness, and surprise. These emotions are understood to be cross-culturally and universally communicated with particular facial expressions.

DEMO: https://azure.microsoft.com/en-us/services/cognitive-services/emotion/

Bing Speech API converts audio to text, understands intent, and converts text back to speech.

DEMO: https://azure.microsoft.com/en-us/services/cognitive-services/speech/

Speaker Recognition API identifies individual speakers.

DEMO: https://azure.microsoft.com/en-us/services/cognitive-services/speaker-recognition/

Text Analytics API detects sentiment, key phrases, topics, and language from your text.

DEMO: https://azure.microsoft.com/en-us/services/cognitive-services/text-analytics/

Google has created a website to showcase AI use cases – https://aiexperiments.withgoogle.com/

Here are a couple of experiments, which will give you a glimpse into what machine learning can do:

Using Machine Learning techniques that analyze the visual features of artworks, X Degrees of Separation finds pathways between any two artifacts, connecting the two through a chain of artworks. This network of connected artworks allows X Degrees of Separation to take us on the scenic route where serendipity is waiting at every step: surprising connections, masterful works by unknown artists or the hidden beauty of mundane objects.

DEMO: https://artsexperiments.withgoogle.com/xdegrees/

This experiment gives you a peek into how machine learning works, by visualizing high-dimensional data. It's available for anyone to try on the web. It is also open-sourced as part of TensorFlow, so that coders can use these visualization techniques to explore their own data.

DEMO: http://projector.tensorflow.org/

Google also has demos of their cognitive applications:

Google Cloud Natural Language API reveals the structure and meaning of text by offering powerful machine learning models in an easy to use REST API. You can use it to extract information about people, places, events and much more, mentioned in text documents, news articles or blog posts. You can use it to understand sentiment about your product on social media or parse intent from customer conversations happening in a call center or a messaging app. You can analyze text uploaded in your request or integrate with your document storage on Google Cloud Storage.

DEMO: https://cloud.google.com/natural-language/

Google Cloud Speech API enables developers to convert audio to text by applying powerful neural network models in an easy to use API. The API recognizes over 80 languages and variants, to support your global user base. You can transcribe the text of users dictating to an application's microphone, enable command-and-control through voice, or transcribe audio files, among many other use cases. Recognize audio uploaded in the request, and integrate with your audio storage on Google Cloud Storage, by using the same technology Google uses to power its own products.

DEMO: https://cloud.google.com/speech/

Cloud Translation API provides a simple programmatic interface for translating an arbitrary string into any supported language. Translation API is highly responsive, so websites and applications can integrate with Translation API for fast, dynamic translation of source text from the source language to a target language (e.g., French to English). Language detection is also available in cases where the source language is unknown. The underlying technology pushes the boundary of Machine Translation and is updated constantly to seamlessly improve translations and introduce new languages and language pairs.

DEMO: https://cloud.google.com/translate/

Google Cloud Vision API enables developers to understand the content of an image by encapsulating powerful machine learning models in an easy to use REST API. It quickly classifies images into thousands of categories (e.g., "sailboat," "lion," "Eiffel Tower"), detects individual objects and faces within images, and finds and reads printed words contained within images. You can build metadata on your image catalog, moderate offensive content, or enable new marketing scenarios through image sentiment analysis. Analyze images uploaded in the request or integrate with your image storage on Google Cloud Storage.

DEMO: https://cloud.google.com/vision/

Google Cloud Video Intelligence API makes videos searchable, and discoverable, by extracting metadata with an easy to use REST API. You can now search every moment of every video file in your catalog and find every occurrence as well as its significance. It quickly annotates videos stored in Google Cloud Storage, and helps you identify key nouns entities of your video, and when they occur within the video. Separate signal from noise, by retrieving relevant information at the video, shot or per frame.

DEMO: https://cloud.google.com/video-intelligence/

Google recently introduced **The Teaching Machine** to demonstrate how Deep Learning works. Using just your webcam, you will see a computer learn in real-time.

DEMO: https://teachablemachine.withgoogle.com/

Amazon wants your business, too, so they have made a number of AI applications available. You will need to sign up for the "free preview" to see them in action, but it's worth the time.

Amazon Lex is a service for building conversational interfaces into any application using voice and text. Lex provides the advanced deep learning functionalities of automatic speech recognition (ASR) for converting speech to text, and natural language understanding (NLU) to recognize the intent of the text, to enable you to build applications with highly engaging user experiences and lifelike conversational interactions. With Amazon Lex, the same deep learning technologies that power Amazon Alexa are now available to any developer, enabling you to quickly and easily build sophisticated, natural language, conversational bots ("chatbots").

DEMO: https://aws.amazon.com/lex/

Amazon Polly is a service that turns text into lifelike speech. Polly lets you create applications that talk, enabling you to build entirely new categories of speech-enabled products. Polly is an Amazon AI service that uses advanced deep learning technologies to synthesize speech that sounds like a human voice. Polly includes 47 lifelike voices spread across 24 languages, so you can select the ideal voice and build speech-enabled applications that work in many different countries.

DEMO: https://aws.amazon.com/polly/

Amazon Rekognition is a service that makes it easy to add image analysis to your applications. With Rekognition, you can detect objects, scenes, and faces in images. You can also search and compare faces. Rekognition's API enables you to quickly add sophisticated deep learning-based visual search and image classification to your applications.

DEMO: https://aws.amazon.com/rekognition/

Amazon Machine Learning is a service that makes it easy for developers of all skill levels to use machine learning technology. Amazon Machine Learning provides visualization tools and wizards that guide you through the process of creating machine learning (ML) models without having to learn complex ML algorithms and technology. Once your models are ready, Amazon

Machine Learning makes it easy to obtain predictions for your application using simple APIs, without having to implement custom prediction generation code, or manage any infrastructure.

DEMO: https://aws.amazon.com/machine-learning/

If you happen to have your college essay handy, or one your kid wrote, try out a cognitive app designed to predict which private schools in New York City are the best fit for your personality, and as a bonus, what careers. I used it with my stepson's essay, and it correctly suggested schools with strong technical and math programs as well as research scientist and engineer for careers. He is a Chemical Engineering major. The essay I used had no mention of his interest in STEM subjects. See how it does with you or yours (you can use any of your writing, and it can be a combination of works) - https://nyc-school-finder.mybluemix.net/

Every day new providers are popping up to offer variations on the services above, and introducing whole new concepts. While the leaders have more money, they are not the sole owners of innovation. Use the descriptions in the demos to search for competitors.

If your organization has implemented Cognitive Computing then see this as a way to evaluate your current technology. If you have not, then this is the time kick the tech tires.

We are moving from single-brand solutions to best-of-breed. This means you will likely be using applications from multiple vendors. Over time you may well switch-out applications much as you do with hardware today. In this environment you want to set up processes to constantly test your current solutions against other offerings.

As you explore always remember AI learns. This means the application you don't like today may well become your favorite tomorrow. Flexibility will be one of the hallmarks of a successful organization moving forward. Gone are the days of decade old technology powering your operations.

I say good riddance!

16

Insight Gallery

"A moment's insight is sometimes worth a life's experience."

—Oliver Wendell Holmes, Sr.

It is natural to ask "What are other people doing?" when considering whether or not to use a new technology or try a new way of doing things. I have put real-world examples throughout the book to ensure this not a theoretical exercise. After all, the future is fun to ponder, but we all need to get things done today.

Here are some more examples how Cognitive Computing is enabling organizations to increase their creativity, reach, and impact:

The Carnegie Museum of Art used the Instagram API to collect photos from across the Pittsburgh metropolitan area and analyzed them for their general mood. This was used to calculate a positive or negative score for the entire city. The score was then used to create a giant data visualization using the lighting on the Gulf Tower. http://www.workergnome.com/work/gulf-tower/

The American Journal of Public Health published "Machine Learning for Social Services: A Study of Prenatal Case Management in Illinois" in April 2017. The study used machine learning to predict high-risk pregnancies. Using data collected by the Illinois Department of Human Services, the model was able to improve on paper-based risk assessment by 36%. http://ajph.aphapublications.org/doi/abs/10.2105/AJPH.2017.303711

The USC Center for Artificial Intelligence in Society (CAIS) was formed in early 2017. "Our motivation was to create a new intellectual space for computer scientists with all these tools we didn't have access to and people in social work science who have this deep understanding of human problems to develop new solutions to vexing, seemingly intractable issues," said Eric Rice, co-founder of the new center. The center is a joint venture between the USC Suzanne Dworak-Peck School of Social Work, where Rice serves as associate professor, and its engineering counterpart, the USC Viterbi School of Engineering, home to center co-founder Milind Tambe, the school's Helen N. and Emmett H. Jones Professor in Engineering. The genesis of the center was a project Tambe and Rice worked on to reduce HIV among homeless adolescents and young adults. They created an algorithm to identify the most influential people within homeless social groups, and then train that person in HIV prevention and encourage friends to get tested. This project resulted in a 60% increase in information sharing in these social networks. https://sowkweb.usc.edu/news/betting-artificial-intelligence-help-humanity

Florida State University psychology researcher, Jessica Ribeiro, PhD, published a paper in the journal of *Clinical Psychological Science*, "Predicting Risk of Suicide Attempts over Time through Machine Learning." The study detailed how machine learning was able to predict suicide attempts with a 80%-90% accuracy as far off as 2 years. The closer to the person's suicide attempt the higher the accuracy, rising to 92% one-week before the attempt when studying general hospital patients.

http://www.socialworktoday.com/news/enews_0417_2.shtml

IBM Watson Health™ - Opioid abuse is among the deadliest population health crises in the United States. In most cases, this stems from a prescription. Understanding the patterns of addiction, learning evidence-based guidelines for responsible prescription, and creating early warning systems are instrumental when battling new addictions. The team will couple advanced machine learning methods with the wealth of Watson Health Truven Health Analytics data to develop insights and make them available to providers, payers

and public health officials to help curb the opioid epidemic.
https://www.ibm.com/blogs/research/2017/08/combatin g-the-opioid-epidemic-with-machine-learning/

St. John's Bread & Life — In 2015, 42.2 million Americans lived in food insecure households. Every day, St. John's Bread & Life, an emergency food provider in New York City, serves more than 2,500 meals with extraordinary efficiency. The team plans to create a cognitive supply chain model of emergency food operations and share it via an API and interactive digital experience. This will allow Bread & Life to share best practices with food providers around the country and provide them with a first-of-a-kind digital outreach campaign to help inform the public about hunger in a way that registers, stays, and changes behaviors.
https://www.breadandlife.org/

In recent years, Wageningen University in the Netherlands and Welgevonden conducted a study that opened many eyes to these new alternatives to protect our rhino friends. What the research found is that prey-animals react differently, depending on the type of threat they encounter, whether it's coming from predators such as a lion or a human nearby, who might be a reserve employee, a tourist, or a poacher. Rhinos are excluded to ensure that their precise whereabouts remain unknown. So how can these rhinos be protected if they are not being tracked? Rather than follow the rhino, all eyes are focused on observing how the prey-animals respond to disturbances, including the presence of potential poachers, versus a tourist or an employee traveling across the reserve in a vehicle. Through IBM's IoT platform, teams monitor and collect sensor-information related to location, movement pattern, direction, and average speed of travel of these animals, and are using this movement and other data to create rule-based patterns, or algorithms, built on the prey-animals' response to perceived threats. https://www.ibm.com/blogs/internet-of-things/protecting-endangered-rhinos/

ICAST - World energy consumption is predicted to grow by 48 percent between 2012 and 2040. Technology solutions only address half the problem of trying to conserve energy. Behavior change is just as important. As multi-family affordable housing is retrofitted for energy conservation, it is also fitted with sensors that allow us to examine behaviors of the building residents to guide and evaluate any interventions that may be undertaken to encourage conservation. The team plans to build a solution for disseminating information about behaviors that reduce energy usage and for measuring the impact of the intervention. http://www.icastusa.org/

The Douglas Mental Health University Institute's Translational Neuroimaging Laboratory at McGill University developed an algorithm to detect signs of dementia before its onset. The algorithm searches brain scans for the buildup of amyloid, a protein that accumulates in the brains of people who develop mild cognitive impairment and, later, dementia. The researchers used PET scans available through the Alzheimer's Disease Neuroimaging Initiative. http://big think.com/articles/scientists-use-machine-learning-to-

spot-alzheimers-dementia-before-onset-of-
symptoms.amp

In no particular order, here are ideas for insight. I have given you some suggestions for quantitative and qualitative data sources, but my hope is you will expand on both and find many more uses cases. I can't wait to see what your organization does to create Big Good.

FOCUS AREA: Event Engagement

USE CASE: Events are typically measured by how many people attended and how much money was raised. What is missing is whether attendees enjoyed the event and whether their attendance delivered higher retention or higher giving in the long-term.

QUANTITATIVE DATA: Constituent demographic data; giving data; event name; event location; event date; event time; event attendance; giving during and immediately after the event.

QUALITATIVE DATA: Did they check-in on FourSquare? Did they tell their friends they went? Did they post pictures from the event? What was the sentiment? Was it a fun event or a boring event? Did they love the speaker or dislike him or her? With your NLP analysis in full operation, be sure to ask post-event, open-ended questions such as, "How did the event make you feel about our organization?"

POTENTIAL INSIGHTS: Impact of event sentiment on giving at the macro and individual level. At the macro level, this will be more useful as you analyze the event year to year. Better look at the impact of the speaker(s) on the attendees. Did shared positive or negative sentiment impact giving outside of the event?

FOCUS AREA: Planned Giving

USE CASE: People often lower, or even stop, giving as they get older yet they leave large gifts to organizations they supported when they were younger. This reality can lead to an organization who relies on RFM to stop communications with a donor because they are lapsed for too long. This is where non-giving engagement becomes critical.

QUANTITATIVE DATA: Attendance information from events, including unable to attend RSVPs; email metrics; responses to surveys; constituent-provided updates to contact information; and social media engagement.

QUALITATIVE DATA: Notes from gift officers indicating reasons for not giving; social media tweets, posts, and comments; verbatim survey responses; communication copy that has been effective with people who made planned gifts; and class notes.

POTENTIAL INSIGHTS: Identify the different journeys people are taking to a planned gift; identify the aspects of the journey that are working the best with different constituent clusters; and identify the right tone to use for communications with these clusters.

FOCUS AREA: Academic Achievement & Student Housing

USE CASE: Look at academic performance and retention as they relate to where students live while attending your school. Examine both on-campus and off-campus locations.

QUANTITATIVE DATA: GPA; class attendance; credits taken; cost; housing location; event attendance; survey ratings; club memberships; and roommate(s).

QUALITATIVE DATA: Social media tweets, posts and reviews; college reviews; teacher notes; counselor notes; and verbatim survey responses.

POTENTIAL INSIGHTS: Determine if there are links between academic performance and student residence; identify high-risk students both in terms of academics and retention; and map the journeys of students who live at different locations to find the paths which lead to positive outcomes.

FOCUS AREA: Student Engagement

USE CASE: Academic performance and retention as they relate to how engaged students are with your institution.

QUANTITATIVE DATA: GPA; class attendance; event attendance; club memberships; counseling activity; survey ratings; and athletic participation.

QUALITATIVE DATA: Social media tweets, comments and posts; verbatim survey responses; teacher notes; and counselor notes.

POTENTIAL INSIGHTS: Develop a solid student engagement score; identify the most effective types of engagement; and discover the interconnections between types of engagement to know the best recommendations to make for students.

FOCUS AREA: Alumni/ae Engagement

USE CASE: Look at how giving, and retention, relate to engagement activities.

QUANTITATIVE DATA: Giving data; prospect management touch points; event attendance including unable to attend RSVPs; survey ratings; social media metrics; and communication metrics.

QUALITATIVE DATA: Gift officer notes; social media tweets, posts, and comments; verbatim survey responses; class notes; and incoming communication content.

POTENTIAL INSIGHTS: A solid alumni engagement score; map of the journeys different alumni/ae clusters are taking; identify alumni/ae who are at high-risk of stopping their giving; and identify alumni/ae who are likely to increase their giving.

FOCUS AREA: Event Pictures

USE CASE: Organizations with events such as walk-a-thons and 5Ks to raise funds can use visual recognition to identify constituents who are associating their brand with yours.

QUANTITATIVE DATA: Event attendee demographic data and giving data.

QUALITATIVE DATA: Pictures from previous events with people wearing your organization's logo; pictures of your logo on wearable items; and pictures of your logo. All of these should show your logo in all the forms it might appear.

POTENTIAL INSIGHTS: A measure of engagement and sentiment is whether people are posting pictures from your event. This can be correlated to giving, retention, and also giving around the event.

FOCUS AREA: Campus Pictures

USE CASE: You want to know when people are posting pictures of your campus which associates themselves with your brand.

QUANTITATIVE DATA: Constituent demographics and giving.

QUALITATIVE DATA: Pictures of the buildings on your campus; stadiums; and any iconic landmarks where people take pictures.

POTENTIAL INSIGHTS: Correlating giving and retention to the association of your constituents' identity with your brand. Tailor your organization's use of pictures on social media with insights regarding what your constituents are posting, sharing, and liking.

FOCUS AREA: Grateful Patient Satisfaction and Giving

USE CASE: Correlating satisfaction with patient care and giving can help inform fundraising as well as potentially help the clinical side improve patient experience.

QUANTITATIVE DATA: Patient satisfaction survey quantitative data elements; donor demographic data; and giving data.

QUALITATIVE DATA: Patient satisfaction survey qualitative data elements (free text); call reports; social media comments; social media posts; and review site posts.

POTENTIAL INSIGHTS: Determine if there is a correlation between overall patient satisfaction and giving; giving level; and percentage of potential. Determine if particular elements of patient satisfaction correlate to giving.

FOCUS AREA: Financial Fraud Detection

USE CASE: Looking for financial activity indicating internal fraud.

QUANTITATIVE DATA: Financial transaction data and constituent demographic and financial data.

QUALITATIVE DATA: Employee job descriptions and Data Governance policies.

POTENTIAL INSIGHTS: Identification of fraudulent financial activity; and identification of people who are accessing and/or exporting financial information without authorization.

FOCUS AREA: Grant Management (grantee)

USE CASE: Analysis of grants throughout the process, from application to impact reporting.

QUANTITATIVE DATA: Financial and demographic data included in grant proposals; active grants; and inactive grants.

QUALITATIVE DATA: Grant proposals; field reports; impact reports; correspondence with grant makers.

POTENTIAL INSIGHTS: Monitoring how well grant results are matching to promised outcomes; sentiment of grant makers towards the organization and/or the grant in particular; and grant journey analytics to identify areas for improvement and areas doing well.

FOCUS AREA: Length of Time a Donor Gives

USE CASE: One of the most powerful metrics you can influence is length of time a donor supports your organization. Understanding why people give longer or stop giving will help you allocate resources to those factors driving long-term relationships.

QUANTITATIVE DATA: Constituent demographics; giving; relationship managers; number of continuous years people have given; all touch points.

QUALITATIVE DATA: Sentiment analysis; call reports; social media; verbatim survey responses; class notes; and interest analysis.

INSIGHT POTENTIAL: Discovering the cluster characteristics of different constituent groups as they relate to retention time; understanding the journey of these clusters throughout your organization; discovering what is influencing donors to continue giving; and discovering what is influencing them to stop.

FOCUS AREA: Grant Proposals (grantee)

USE CASE: Discover what makes grant proposals successful and what is causing them to not be funded.

QUANTITATIVE DATA: Grant proposal basic data; dates; ask; amount received.

QUALITATIVE DATA: Successful grant proposals; unsuccessful grant proposals; project and program details; and relevant grant maker giving guidelines.

POTENTIAL INSIGHTS: How well grant proposals are capturing the impact of the projects and programs for which funding is being sought; how well grant proposals are aligned with the giving priorities of the grant makers from which they are seeking funding.

FOCUS AREA: Student Roommate Matching

USE CASE: Match roommates based on personality insights.

QUANTITATIVE DATA: Past successful and unsuccessful roommate demographics and academics for roommates.

QUALITATIVE DATA: Past successful and unsuccessful roommate-authored content including applicable essays and social media posts by roommates.

POTENTIAL INSIGHTS: Identify emotion (Anger; Disgust; Fear; Joy; and Sadness); Language Style (Analytical; Confident; Tentative); and Social Tendencies (Openness; Conscientiousness; Extraversion; Agreeableness; and Emotional Range). Identify the personalities of the roommates who lived well together.

FOCUS AREA: Office of Research & Commercialization

USE CASE: Match research being done internally to companies needing to license the research.

QUANTITATIVE DATA: Basic data about research projects and patents; SIC and NAICS codes for relevant companies.

QUALITATIVE DATA: Detailed descriptions of the research projects and patents; descriptions of relevant companies.

POTENTIAL INSIGHTS: Identify companies in need of your institution's research and patents; identify research and patents having no or a very limited market.

FOCUS AREA: Email Subject Line

USE CASE: Create subject lines to increase open rates by better matching constituent interests as well as word choice.

QUALITATIVE DATA: Constituent demographics; email opens; clicks; forwards

QUALITATIVE DATA: Subject line text; constituent interest data.

POTENTIAL INSIGHTS: Correlation of subject line content and past email actions identify where you are in sync with constituents; identification of email subjects to match constituent interests.

FOCUS AREA: Marketing Channels

USE CASE: Identify the best channel(s) for constituent by looking at actions across all channels.

QUANTITATIVE DATA: Constituent demographics; constituent giving; all channel structured data.

QUALITATIVE DATA: Written communications about channels; social media posts.

POTENTIAL INSIGHTS: Factors driving the success or failure within each channel. Factors driving the success or failure for multi-channel marketing efforts. How channels influence the results of each other.

FOCUS AREA: Marketing Copy Tone

USE CASE: Identify the tone of your marketing copy to make sure it is consistent and effective.

QUANTITATIVE DATA: Constituent demographics; giving; marketing structured data including campaign codes; opens; clicks, etc.

QUALITATIVE DATA: Marketing copy for particular campaigns across channels.

POTENTIAL INSIGHTS: Identify emotion (Anger; Disgust; Fear; Joy; and Sadness); Language Style (Analytical; Confident; Tentative); and Social Tendencies (Openness; Conscientiousness; Extraversion; Agreeableness; and Emotional Range); use this analysis to see what combination of factors are leading to marketing copy that is most effective with different constituent clusters.

FOCUS AREA: Caller Tone Analysis

USE CASE: Evaluate how your callers are interacting with your mission's constituents.

QUANTITATIVE DATA: Caller demographics; call results; call-structured details.

QUALITATIVE DATA: Transcripts of calls with caller identified.

INSIGHT POTENTIAL: Identify emotion (Anger; Disgust; Fear; Joy; and Sadness); Language Style (Analytical; Confident; Tentative); and Social Tendencies (Openness; Conscientiousness; Extraversion; Agreeableness; and Emotional Range).

You can use this analysis in combination with call results to see which callers are using the right tone to garner desired results. Match callers to constituent clusters which might be driven by age, gender, and/or geography.

FOCUS AREA: Stewardship Communications

QUANTITATIVE DATA: Constituent demographics; giving including number of years of continuous giving; stewardship communication structured data.

QUALITATIVE DATA: Full copy of stewardship communications; emails; letters; brochures; website; and social media.

POTENTIAL INSIGHTS: Identify emotion (Anger; Disgust; Fear; Joy; and Sadness); Language Style (Analytical; Confident; Tentative); and Social Tendencies (Openness; Conscientiousness; Extraversion; Agreeableness; and Emotional Range). You can use this analysis to see what copy is most effective in increasing retention rates and giving levels.

FOCUS AREA: Admissions Personality Analysis

USE CASE: Determine the personality traits of successful and unsuccessful students to help identify high potential and at-risk admission candidates.

QUANTITATIVE DATA: Demographic data on graduates and dropouts; academic details on both groups.

QUALITATIVE DATA: Admissions essay answers; social media posts; other written content.

POTENTIAL INSIGHTS: Identify 52 personality traits including needs; values; agreeableness; openness; conscientiousness; extraversion; and emotional range of graduate clusters and dropout clusters. You can use this to find candidates who are more likely to be successful at your institution. You can identify new students who may need particular types of outreach to be successful.

FOCUS AREA: Staff Personality Analysis

USE CASE: Understand your staff's personalities to better match them to activities as well as building better teams.

QUANTITATIVE DATA: Staff demographics; staff performance structured data.

QUALITATIVE DATA: All available staff-authored content including proposals; emails; social media posts; and verbatim answers to survey questions.

POTENTIAL INSIGHTS: Gain a better understanding of the strengths of your team and how they match up with your constituency. This can also be used for hiring.

FOCUS AREA: Stewardship Communications Center

USE CASE: Create a self-serve stewardship center where a donor or prospective donor can learn about the impact of your mission.

QUANTITATIVE DATA: Relevant structured data about your projects and programs such as costs; number of people served, etc.

QUALITATIVE DATA: All written material about your projects and programs from websites; marketing materials; and impact reports.

POTENTIAL INSIGHTS: By establishing a self-serve impact center, this enables your organization to provide 24/7/365 answers for donors and prospective donors. You will learn what your constituents are interested in as they ask questions. These insights can be integrated with your interest analysis. These insights can also guide marketing communications, as you discover areas of particular interest.

FOCUS AREA: Career Matching

USE CASE: Match students to careers and internships based on a combination of major(s); academic performance; and personality.

QUANTITATIVE DATA: Student academic data; student demographic data.

QUALITATIVE DATA: Student-authored content including college essay; written reports; and social media posts.

POTENTIAL INSIGHTS: Identify 52 personality traits including needs; values; agreeableness; openness; conscientiousness; extraversion; and emotional range of graduates who have gone on to be successful within different careers. Use to identify the best careers for current students.

FOCUS AREA: Volunteers

USE CASE: Identify volunteers based on their personalities, interests, and engagement with your organization.

QUANTITATIVE DATA: Constituent demographics; giving; and touch points.

QUALITATIVE DATA: Authored content from emails and social media. Interests and involvement as volunteers with other organizations from social media.

INSIGHT POTENTIAL: Understand the personality types that are successful as volunteers for your organization; identify people with those personality types; and match volunteers to particular opportunities based on a combination of personality and interests.

FOCUS AREA: Mission Program Analysis

USE CASE: Create journey maps for different aspects of your mission.

QUANTITIVE DATA: Touch point metrics such as receiving a benefit; quantifiable outcomes such as employment; survey ratings; outside ratings and rankings; and critical components of the program such as supplies and employees.

QUALITATIVE DATA: Beneficiary verbatim survey responses; mission employee notes and verbatim survey responses; volunteer notes and verbatim survey responses; external reviews; and social media tweets, posts, and comments.

POTENTIAL INSIGHTS: Identify strong and weak aspects of your programs; identify beneficiaries who are at high risk for failure; identify the best journeys for different types of beneficiaries to take.

FOCUS AREA: Fundraising Economic Analysis

USE CASE: Determine the correlation between economic data and your fundraising results.

QUANTITATIVE DATA: Federal Reserve Economic Data (FRED); Philadelphia Federal Reserve Coincident Index; relevant State economic data; relevant country economic data; IRS giving statistics for relevant geographic locations.

QUALITATIVE DATA: National economic reports; relevant State and local economic reports; relevant country reports; and reports on giving.

POTENTIAL INSIGHTS: Better understand how your fundraising results are lining up with the economic conditions of the locations where your constituents reside; identify high-potential locations; and identify low-potential locations.

FOCUS AREA: College Athletics and Fundraising

USE CASE: Determine the correlation between athletic event attendance and giving to schools.

QUANTITATIVE DATA: Athletic ticketing information; demographic data; survey ratings; and giving data.

QUALITATIVE DATA: Social media tweets, posts, and comments; verbatim survey responses; and gift officer reports.

POTENTIAL INSIGHTS: Identify sport affinities; create engagement scores for each sport; identify athletic event attendance's impact on giving; and identify potential connections between alumni/ae based on shared event attendance.

FOCUS AREA: College Performing & Fine Arts & Fundraising

USE CASE: Determine the correlation between attending school sponsored performing and fine arts events and giving.

QUANTITATIVE DATA: Event ticketing information; demographic data; giving data; survey ratings.

QUALITATIVE DATA: Social media tweets, posts, and comments; verbatim survey responses; and gift officer reports.

POTENTIAL INSIGHTS: Identify interests; create engagement scores for performing and fine arts; identify attendance's impact on giving; identify potential connections between alumni/ae based on shared event attendance

FOCUS AREA: Social Network Engagement & Fundraising

USE CASE: Determine the social media engagement level of your constituents.

QUANTITATIVE DATA: All of your organization's social networks; demographic data; giving data; and social network metrics.

QUALITATIVE DATA: Social media tweets, posts, and comments; gift officer reports.

POTENTIAL INSIGHTS: Identify which of your social networks within which your constituents are engaging; create engagement scores for each of your social networks; discover interests and sentiment; and determine the correlation between social engagement and giving.

FOCUS AREA: Mission Ontology

USE CASE: Build ontology for your mission to create an interest matrix to connect your constituents to your projects and programs.

QUANTITATIVE DATA: N/A

QUALITATIVE DATA: Website(s); cases for support; grant proposals; marketing materials; white papers; and field reports.

POTENTIAL INSIGHTS: A complete ontology of your organization's mission; the ability to map your mission to the interests of your constituents.

FOCUS AREA: Mission Area Analysis

USE CASE: Understand how your constituency perceives aspects of your mission.

QUANTITATIVE DATA: Mission area demographics; mission area metrics; survey ratings; external assessment metrics; and social media metrics.

QUALITATIVE DATA: Social media tweets, posts, and comments; verbatim survey responses; gift officer reports; and external reports.

POTENTIAL INSIGHTS: Identify sentiment about mission areas; identify sentiment over time about mission areas; identify what is driving positive or negative sentiment; and determine the correlation between sentiment and giving to the mission areas.

FOCUS AREA: Major Gift Optimization

USE CASE: Understanding how effective your fundraising program is with different types of constituents (entrepreneur; professional; old-money, etc.) and constituent clusters.

QUANTITATIVE DATA: Demographic data; professional data; giving data; prospect management data; wealth ratings; and survey ratings.

QUALITATIVE DATA: Gift officer reports; social media tweets, comments, and posts; and verbatim survey responses.

POTENTIAL INSIGHTS: Determine current wealth penetration levels (giving vs. potential) for each type of constituent and each constituent cluster; identify giving patterns for each type of constituent and constituent cluster; map the journey for each type of constituent and constituent cluster; and identify gift officers who are most effective (or not) with certain types of constituents and constituent clusters.

FOCUS AREA: Reputation

USE CASE: Understanding how constituents, and prospective constituents, perceive your organization.

QUANTITATIVE DATA: Survey ratings; external review ratings.

QUALITATIVE DATA: Gift officer reports; verbatim survey responses; verbatim external reviews; social media tweets, comments, and posts; industry reports; and news articles.

POTENTIAL INSIGHTS: Know the opinions and feelings of your constituents about your organization overall as well as specific aspects of it; develop a Reputation Score; and develop scores for different aspects of your organization.

FOCUS AREA: Channel Effectiveness

USE CASE: Determine the effectiveness of each channel.

QUANTITATIVE DATA: Channel metrics; giving broken out by channel; and industry benchmark metrics for each channel.

QUALITATIVE DATA: Gift officer reports; call center notes; call center audio for speech-to-text; incoming communications; and social media tweets, posts, and comments.

POTENTIAL INSIGHTS: Map of your constituents' interactions and giving by channel; correlations between channels; correlation between channel activity and giving; and channel scores.

FOCUS AREA: Email Content

USE CASE: Understand the correlation between content and results.

QUANTITATIVE DATA: Email metrics; demographic data; giving data.

QUALITATIVE DATA: Text of emails; subject line text; content linked to; images; video.

POTENTIAL INSIGHTS: Identify words, phrases, tone, emotion, images, videos, and mission content effective with different constituents and constituent clusters; enable cognitive creation of future content.

FOCUS AREA: Event Effectiveness

USE CASE: Understanding the impact of events on giving, engagement, and retention; and survey ratings.

QUANTITATIVE DATA: Event metrics; giving data; demographic data; and any IoT data.

QUALITATIVE DATA: Event copy, images, video, and mission content; website(s); marketing materials; verbatim survey responses; pictures; videos.

POTENTIAL INSIGHTS: Understanding the correlations between event attendance and giving, engagement, and retention; identifying effective words, phrases, tone, emotion, images, videos, and mission content; and creating event scores.

FOCUS AREA: Direct Mail Drivers

USE CASE: Discover what is influencing your direct mail results.

QUANTITATIVE DATA: Direct mail metrics; giving data; demographic data; weblogs; weather; economic data; and other channel metrics.

QUALITATIVE DATA: Direct mail copy, images and mission content; social media tweets, comments, and posts; gift officer reports; and incoming communications.

POTENTIAL INSIGHTS: Discover the correlations between direct mail and other channels; discover correlations between weather and economic data and your direct mail results; discover correlations between copy, images, and mission content and direct mail results.

FOCUS AREA: Crowdfunding

USE CASE: Determining new giving interests from crowdfunding giving.

QUANTITATIVE DATA: Crowdfunding metrics; giving data; and demographic data.

QUALITATIVE DATA: Crowdfunding project and funding websites, copy, images, and videos; gift officer reports; all other projects and programs of the organization.

POTENTIAL INSIGHTS: Map all crowdfunding projects and programs to applicable projects and programs in your main fundraising program; and determine new and current interests of constituents.

FOCUS AREA: Campaign Videos

USE CASE: Creating more effective campaign videos.

QUANTITATIVE DATA: Campaign video metrics.

QUALITATIVE DATA: Previous campaign videos; video about your organization; and videos you like from other organizations.

POTENTIAL INSIGHTS: Identify the best video footage to use in your campaign videos.

FOCUS AREA: Professor Assessment

USE CASE: Evaluate perceptions of professor by their current and former students.

QUANTITATIVE DATA: Student academic information; student survey ratings; and school ratings.

QUALITATIVE DATA: Students' verbatim class reviews; college and professor review sites; and social media tweets, posts, and comments.

POTENTIAL INSIGHTS: Discover correlations between types of professors and academic performance; identify areas for improvement or reinforcement; and create a professor score to help better match student personality types with professors.

FOCUS AREA: Student Success

USE CASE: Find paths for students to succeed.

QUANTITATIVE DATA: Academic information; demographics; housing data; club memberships; event attendance; school activity involvement; counseling metrics; financial aid; student review ratings; roommate(s) information; and social media metrics.

QUALITATIVE DATA: Counselor notes; professor notes; student verbatim reviews; social media tweets, posts, and comments.

POTENTIAL INSIGHTS: Map the journey of students and student clusters; identify areas for improvement or reinforcement; and identify high-risk students for academic problems, dropping, and/or transferring.

FOCUS AREA: Student Career Counseling

USE CASE: Improve matching of students to career opportunities.

QUANTITATIVE DATA: Demographic data; academic data; jobs; internships; academic research positions; alumni/ae employment information; alumni/ae academic information; salary information by position/industry.

QUALITATIVE DATA: Authored content for personality analysis; social media tweets, posts, and comments.

POTENTIAL INSIGHTS: Discover the best academic paths for students seeking particular careers; match students to the best potential career; match students to the best potential employers; and match students with alumni/ae mentors.

17

So You Want to Learn More

"Live as if you were to die tomorrow. Learn as if you were to
live forever."

—Mahatma
Gandhi

For those of you with a good foundational knowledge of
statistics and/or computer science and some coding skills, I
suggest you download open-source machine and deep
learning libraries from Python and TensorFlow. There is plenty
of data to play with at www.Data.gov or www.Kaggle.com. It
will not be long before you are ready to ask your organization
for some data to begin turning it into actionable insights.

If you want to learn either from the ground-up or about
particular topics here are some free and fee-based sources to
build your knowledge and your skills:

CognitiveClass AI

Free Data Science and Cognitive Computing Courses

https://cognitiveclass.ai/

Practical Machine Learning Tutorial with Python Introduction

The objective of this course is to give you a holistic understanding of machine learning, covering theory, application, and inner workings of supervised, unsupervised, and deep learning algorithms. In this series, they cover linear regression, K Nearest Neighbors, Support Vector Machines (SVM), flat clustering, hierarchical clustering, and neural networks.

https://pythonprogramming.net/machine-learning-tutorial-python-introduction/

Dataquest (free and fee-based levels)

Their hands-on method teaches you all the skills you need to become a data scientist, data analyst, or data engineer. Learn by writing code, working with data, and building projects in your browser.

https://www.dataquest.io/

DataCamp (free and fee-based levels)

From the basecamp to the pinnacle of your journey, DataCamp is the first and foremost leader in Data Science Education, offering skill-based training, pioneering technical innovation, and courses from the world's best educators.

https://www.datacamp.com

Udacity

Data Science

Learn data science from industry experts at Facebook, Cloudera, MongoDB, Georgia Tech, and more. They offer courses in data science where you'll learn to solve data-rich problems and apply this knowledge to your Big Data needs. Whether you're new to the field or looking for additional training, they have introductory, advanced, and industry-specific courses to meet your learning goals.

https://www.udacity.com/courses/data-science

Coursera

Ask the right questions, manipulate data sets, and create visualizations to communicate results.

This Specialization covers the concepts and tools you'll need throughout the entire data science pipeline, from asking the right kinds of questions to making inferences and publishing results. In the final Capstone Project, you'll apply the skills learned by building a data product using real-world data. At completion, students will have a portfolio demonstrating their mastery of the material.

https://www.coursera.org/specializations/jhu-data-science

Code Academy

In this course, you'll use Python to interact with the Twitter API and IBM Watson's Personality Insights API in order to analyze traits shared between two Twitter users.

https://www.codecademy.com/learn/ibm-watson

There are also an increasing number of degree and certificate programs offered by leading universities including Harvard, MIT, Stanford, Georgia Tech, and the University of Central Florida. You will find on-line, on-site, and hybrid programs to fit your needs, budget, and your schedule.

No matter what your level is now, I encourage you to learn about Cognitive Computing. Ignorance is not an option going forward. You don't have to learn to code or the intricacies of how it all works, but you do need to be able to understand its impact on your organization.

What you do is too important to waste time fighting the future with tired excuses such as "it will never happen here" or "that's for other organizations, not ours."

Get in the race, and run it like your mission depended on it, because it does.

"The most effective way to do it is to do it."

—Amelia Earhart

About the Author

David M. Lawson

David has been a leader in bringing technology and insights to the philanthropic community for a generation. He created the first asset-based wealth screening service as well as the first software to manage screening data. In 1997, he founded Prospect Information Network (P!N), which became the largest wealth screening company before being purchased in 2004. P!N received the InfoCommerce Model of Excellence Award and introduced the first Software-as-a-Service application to support fundraising analytics.

David is CEO and co-founder of NewSci, LLC. In 2014, he was among the early application developers approved to use IBM Watson. This work and subsequent projects encompassing multiple aspects of Big Data and Cognitive Computing led David to write *Big Good: Philanthropy in the Age of Big Data & Cognitive Computing*.

David is the recipient of the Apra Distinguished Service Award and the CASE Crystal Apple Award for Teaching Excellence. He is an Advisory Board Member of the University of Central Florida Master of Science in Data Analytics program. David is a contributing author to *People to People Fundraising: Social Networking and Web 2.0 for Charities*, and *Major Donors: Finding Big Gifts in Your Database and Online*. A frequent speaker, David has presented at numerous conferences including the AFP International Conference, AASP Summit, AHP International Conference, Cloud Expo, Apra International Conference, and CASE. He is also a co-founder of Domi Station, a business incubator in Tallahassee, Florida, and a co-founder of WorkingPhilanthropy.com, LLC. David lives with his wife (and Editor), Lori, in Tallahassee.

www.ingramcontent.com/pod-product-compliance
Lightning Source LLC
Chambersburg PA
CBHW061128220326
41599CB00024B/4207